GIS TUTORIAL for ArcGIS Pro 3.4

GIS TUTORIAL for ArcGIS Pro 3.4

WILPEN L. GORR
KRISTEN S. KURLAND

Esri Press
REDLANDS | CALIFORNIA

Esri Press, 380 New York Street, Redlands, California 92373-8100
Copyright © 2025 Esri
All rights reserved.
Printed in the United States of America
29 28 27 26 25   1 2 3 4 5 6 7 8 9 10

Library of Congress Control Number: 2024952128
ISBN: 9781589488151

The information contained in this document is the exclusive property of Esri or its licensors. This work is protected under United States copyright law and other international copyright treaties and conventions. No part of this work may be reproduced or transmitted in any form or by any means, electronic or mechanical, including photocopying and recording, or by any information storage or retrieval system, except as expressly permitted in writing by Esri. All requests should be sent to Attention: Director, Contracts and Legal Department, Esri, 380 New York Street, Redlands, California 92373-8100, USA.

The information contained in this document is subject to change without notice.

US Government Restricted/Limited Rights: Any software, documentation, and/or data delivered hereunder is subject to the terms of the License Agreement. The commercial license rights in the License Agreement strictly govern Licensee's use, reproduction, or disclosure of the software, data, and documentation. In no event shall the US Government acquire greater than RESTRICTED/LIMITED RIGHTS. At a minimum, use, duplication, or disclosure by the US Government is subject to restrictions as set forth in FAR §52.227-14 Alternates I, II, and III (DEC 2007); FAR §52.227-19(b) (DEC 2007) and/or FAR §12.211/12.212 (Commercial Technical Data/Computer Software); and DFARS §252.227-7015 (DEC 2011) (Technical Data - Commercial Items) and/or DFARS §227.7202 (Commercial Computer Software and Commercial Computer Software Documentation), as applicable. Contractor/Manufacturer is Esri, 380 New York Street, Redlands, California 92373-8100, USA.

Esri products or services referenced in this publication are trademarks, service marks, or registered marks of Esri in the United States, the European Community, or certain other jurisdictions. To learn more about Esri marks, go to links.esri.com/EsriProductNamingGuide. Other companies and products or services mentioned herein may be trademarks, service marks, or registered marks of their respective mark owners.

For purchasing and distribution options (both domestic and international), please visit esripress.esri.com.

Cover credit: WorldClim, Esri

Contents

Preface··ix
Acknowledgments··xi
About this book···xiii

Part 1 Using, making, and sharing maps ···························· 1

Chapter 1 Introducing ArcGIS ·· 3
Tutorial 1-1: Getting an overview of ArcGIS Pro································ 4
Tutorial 1-2: Navigating ArcGIS Pro·· 13
Tutorial 1-3: Working with attribute data······································ 20
Tutorial 1-4: Symbolizing maps·· 27

Chapter 2 Map design ·· 33
Tutorial 2-1: Symbolizing qualitative attributes······························· 34
Tutorial 2-2: Labeling features and configuring pop-ups······················ 37
Tutorial 2-3: Filtering with definition queries································ 41
Tutorial 2-4: Creating choropleth maps for quantitative attributes············ 45
Tutorial 2-5: Displaying data using graduated and proportional point symbols···· 49
Tutorial 2-6: Normalizing population maps with custom scales················ 51
Tutorial 2-7: Creating dot density maps······································· 56
Tutorial 2-8: Setting visibility ranges·· 58

Chapter 3 Maps for end users ······································ 63
Tutorial 3-1: Building layouts and charts······································ 64
Tutorial 3-2: Sharing maps online··· 71
Tutorial 3-3: Creating a story and a briefing in ArcGIS StoryMaps············ 77
Tutorial 3-4: Creating a dashboard in ArcGIS Dashboards····················· 85

Part 2 Working with spatial data · 93

Chapter 4 File geodatabases · 95

Tutorial 4-1: Importing data into a new ArcGIS Pro project · · · · · · · · · · · · · · · · · · 96
Tutorial 4-2: Modifying attribute tables · 101
Tutorial 4-3: Carrying out attribute queries · 109
Tutorial 4-4: Aggregating data with spatial joins ·119
Tutorial 4-5: Using central point features for polygons · · · · · · · · · · · · · · · · · · · 120
Tutorial 4-6: Creating a new table for a one-to-many join · · · · · · · · · · · · · · · · · 122

Chapter 5 Spatial data · 125

Tutorial 5-1: Working with world map projections · 126
Tutorial 5-2: Working with US map projections · 129
Tutorial 5-3: Setting projected coordinate systems · 130
Tutorial 5-4: Working with vector data formats · 134
Tutorial 5-5: Working with US Census map layers and data tables · · · · · · · · · · · · 138
Tutorial 5-6: Downloading geospatial data · 145

Chapter 6 Geoprocessing · 151

Tutorial 6-1: Dissolving features to create neighborhoods and fire divisions
 and battalions · 152
Tutorial 6-2: Extracting and clipping features for a study area · · · · · · · · · · · · · · 155
Tutorial 6-3: Merging water features · 159
Tutorial 6-4: Appending firehouses and police stations to EMS facilities · · · · · · · · 160
Tutorial 6-5: Intersecting features to determine streets in fire company zones · · · 162
Tutorial 6-6: Using Union on neighborhoods and land-use features · · · · · · · · · · · 165
Tutorial 6-7: Using the Tabulate Intersection tool · 169

Chapter 7 Digitizing · 173

Tutorial 7-1: Editing polygon features · 174
Tutorial 7-2: Creating and deleting polygon features · 182
Tutorial 7-3: Using cartography tools · 187
Tutorial 7-4: Transforming features · 189

Chapter 8 Geocoding · 197

Tutorial 8-1: Geocoding data using zip codes · 199
Tutorial 8-2: Geocoding street addresses · 205

Part 3 Applying advanced GIS technologies · · · · · · · · · · · · · · · 211

Chapter 9 Spatial analysis · 213

Tutorial 9-1: Using buffers for proximity analysis · 214

Tutorial 9-2: Using multiple-ring buffers · 216

Tutorial 9-3: Creating multiple-ring service areas for calibrating a gravity model · · · · 219

Tutorial 9-4: Using Network Analyst to locate facilities · 227

Tutorial 9-5: Performing data cluster analysis · 232

Chapter 10 Raster GIS · 239

Tutorial 10-1: Processing raster datasets · 240

Tutorial 10-2: Making a kernel density (heat) map · 247

Tutorial 10-3: Building a risk index model · 253

Chapter 11 3D GIS · 269

Tutorial 11-1: Exploring a global scene · 270

Tutorial 11-2: Creating a local scene and TIN surface · 273

Tutorial 11-3: Creating z-enabled features · 277

Tutorial 11-4: Creating features and line-of-sight analysis using lidar data · · · · · · · · 280

Tutorial 11-5: Working with 3D features · 289

Tutorial 11-6: Using procedural rules and multipatch models · · · · · · · · · · · · · · · · · 294

Tutorial 11-7: Creating an animation · 298

Data source credits · 303

Preface

Welcome to *GIS Tutorial for ArcGIS® Pro 3.4*. This sixth-edition, step-by-step workbook focuses on ArcGIS Pro but also covers ArcGIS Online, ArcGIS StoryMaps℠, and other ArcGIS software.

ArcGIS Pro is Esri's premier desktop GIS application, with an intuitive user interface for creating, using, and analyzing professional 2D and 3D maps. ArcGIS ModelBuilder™ allows you to build models that automate multiple geoprocessing steps, all without computer coding.

ArcGIS Online is software as a service (SaaS) for making, analyzing, and sharing interactive maps. You can easily publish maps authored in ArcGIS Pro to ArcGIS Online and integrate map layers from ArcGIS Living Atlas of the World. Curated by Esri®, ArcGIS Living Atlas is the world's foremost collection of ready-to-use layers and maps. With your finished maps stored in ArcGIS Online, you can easily share them in ArcGIS Online apps. ArcGIS StoryMaps allows you to incorporate text, charts, and other content with interactive maps to tell a story or make a report. ArcGIS Dashboards allows you to follow data patterns in maps over time.

This book is a complete learning system for GIS, including content from more than 25 years of teaching GIS using Esri products and tutorials. Students successfully use our books in and out of classrooms. We have taught high school students, career professionals, undergraduate students, graduate students, and distance-learning students across many disciplines. We teach using a combination of lectures and lab sessions and always include student-designed GIS projects as a final requirement in our courses. Our interactions with students are important sources of ideas and feedback for our books.

Attention, self-learners: Buying this book and gaining access to its accompanying materials is like taking a GIS class. You will receive the step-by-step tutorial book, free one-year access to ArcGIS Pro software, tutorial data, end-of-chapter assignments, and lecture slides and video lectures on underlying concepts for each chapter. See the "About This Book" section of the book.

This book includes wide-ranging real-world data and GIS applications that require the integration of different GIS processing steps and workflows to address realistic problems, such as the following:

- Accessing urgent care health clinics by low-income populations
- Analyzing employment prospects in arts fields across US cities
- Digitizing buildings from satellite images and 3D modeling buildings using lidar data for city planning
- Locating defibrillators to resuscitate heart attack victims
- Data-mining crime patterns in a city for insights into criminal behavior
- Managing nonemergency 311 calls for government services in a city
- Building a multiple-criteria index of poverty in a city

Important concepts are presented at the start of each chapter or in tutorial introductions. Sections on concepts are separated from tutorial steps to ensure continuous, hands-on computer work. Tutorials are illustrated with software interface images and comments. We try to keep these sections concise to help you stay focused. The tutorials are divided into subsections to facilitate learning and retention and provide a sense of accomplishment. In Your Turn assignments within tutorials, you will repeat steps you just learned but in slightly modified ways. These short assignments help you internalize steps and workflows. End-of-chapter assignments, available online, further the objective of internalizing the material.

This book is organized in a sequence that best motivates and facilitates your learning of GIS. In part 1, you will work with and modify existing ArcGIS maps. You will learn to use the ArcGIS Pro interface to navigate and query maps and underlying data, symbolize a range of map types using cartographic principles, build map layouts, use ArcGIS StoryMaps to present your results, and build a dashboard to monitor data.

In part 2, you will delve more deeply into the use of real data—finding, understanding, and preparing spatial data for use. You will download spatial data from public repositories, store and process spatial data in file geodatabases, modify spatial data with geoprocessing tools for mapping, digitize your own spatial data, and geocode tabular data for mapping.

In part 3, you will focus on analyzing spatial data and maps to solve problems. You will apply several unique GIS methods for analyzing spatial relationships, including buffers, service areas, facility location, and data clustering. You will use raster GIS and ModelBuilder to build a tool for analyzing demand for services. Finally, you will work with 3D GIS in the context of city planning.

Acknowledgments

We would like to thank all who made this book possible.

We have taught GIS courses at Carnegie Mellon University since the late 1980s, always using our own lab materials. With the feedback and encouragement of students, teaching assistants, and colleagues, we eventually wrote what became the GIS Tutorial series of workbooks, leading to this book. We are forever grateful for that support.

Faculty members of other universities who have taught GIS using our books have also provided valuable suggestions and feedback. They include Luke Ward of Rocky Mountain College, Irene Rubinstein of Seneca College, An Lewis of the University of Pittsburgh, George Tita of the University of California, Irvine, Walter Witschey of Longwood University, Jerry Bartz of Brookhaven College, and James Querry of Philadelphia University.

We are grateful to the many public servants and vendors who have generously supplied us with interesting GIS applications and data, including Eli Thomas of the Allegheny County Division of Computer Services; Kevin Ford of Campus Design and Facility Development, Carnegie Mellon University; Barb Kviz of the Green Practices program, Carnegie Mellon University; physicians at UPMC Children's Hospital of Pittsburgh; Erol Yildirim, Council for Community and Economic Research; many staff members of the Department of City Planning, City of Pittsburgh; staff members of the New York City Department of City Planning; Wendy Urbanic, Pittsburgh 311 Response Center; Bob Gradeck, Western Pennsylvania Regional Data Center; Michael Radley of the Pittsburgh Citiparks Department; Pat Clark and Traci Jackson of Jackson Clark Partners; Maurie Kelly of Pennsylvania Spatial Data Access (PASDA); staff of the Pennsylvania Resources Council; Kirk Brethauer of the Southwestern Pennsylvania Commission; Steve Benner of Pictometry International Corporation; and employees of several spatial data vendors, including Esri, HERE Technologies, the National Geospatial-Intelligence Agency, US Geological Survey, and National Park Service.

Many technical and expert GIS staff members of Esri reviewed the first draft of this book, and we are grateful for their comments, corrections, and clarifications. It was a pleasure working with these dedicated and talented professionals. Any remaining errors are ours. Finally, we are much indebted to the wonderful staff at Esri Press for their editorial expertise, beautiful design work, efficient production, and distribution of our book.

About this book

GIS Tutorial for ArcGIS Pro 3.4 has been tested for compatibility with ArcGIS Pro 3.4 software and is designed for students in a classroom or others who want to learn GIS on their own. No prior GIS knowledge or experience is needed.

Software requirements and licensing

To do the tutorials in this book, you will need the following: ArcGIS Pro 3.4 installed on a computer that's running the Windows operating system, an internet connection, Microsoft Excel, and a web browser to access ArcGIS Online. The ArcGIS Spatial Analyst™ and ArcGIS Network Analyst™ extensions are required for several tutorials. The ArcGIS student use license provided with the book has these extensions available for your use.

Earlier software versions may not be fully compatible with this tutorial data and may not operate as described in the tutorials. Hardware requirements for ArcGIS Pro are available at links.esri.com/SysReqs.

Information on software trial options, as well as Personal Use and Student Use licensing, can be found at esri.com.

Licensing the software

- Existing credentials: If you have credentials from your institution or organization, you may use these credentials and proceed.
- Student-use license (available for United States users):
    a. Print textbooks (purchased in the United States) come with a code printed inside the back cover.
    b. E-books purchased through VitalSource and labeled as courseware come with a license. After purchase, a code is provided. Visit links.esri.com/BookCode for help locating this code.
    c. Some chapters may require additional products not included in this license. License activation instructions are provided at links.esri.com/GIST3.4License.

Installing the tutorial data

The tutorial data for this book is available at links.esri.com/GISTforPro3.4Data. Download the tutorial data and extract it to your local (C:) drive.

Downloadable data that accompanies this book is covered by a license agreement that stipulates the terms of use. Review this agreement at links.esri.com/LicenseAgreement.

Resources, feedback, and updates

End-of-chapter assignments and data are available in ArcGIS Online (links.esri.com/GISTforPro3.4Assignments). These assignments apply each chapter's concepts and workflows to scenarios and provide more in-depth learning. Instructors often use these resources as additional course content that can be selectively assigned and turned in for grading.

Links to video lectures and lecture slides, which may be useful for instructors or self-learners, can be found on the book's web page at links.esri.com/GISTforPro3.4.

The ArcGIS Pro Help documentation provides comprehensive descriptions of software concepts and tools at links.esri.com/Help.

Feedback, updates, and collaboration are available at Esri Community, the global community of Esri users. Post any questions about this book at links.esri.com/EsriPressCommunity.

GIS is our favorite subject to teach and a favorite class for our students to take. If you have any questions or feedback, you can reach us through email at gorr@cmu.edu or kurland@cmu.edu.

# PART 1

Using, making, and sharing maps

CHAPTER 1

Introducing ArcGIS

LEARNING GOALS

- Get an introduction to ArcGIS®.
- Get an introduction to the ArcGIS Pro user interface.
- Learn to navigate maps.
- Work with tables of attribute data.
- Get an introduction to symbolizing and labeling maps.
- Work with 2D and 3D maps.

Introduction

ArcGIS is an integrated collection of geographic information system (GIS) software developed by Esri® that works seamlessly across desktop computers, the internet, and mobile devices. The tutorials in this first chapter will familiarize you with a major component of this suite: ArcGIS Pro. ArcGIS Pro is a 64-bit desktop GIS application that uses a ribbon interface for 2D and 3D map authoring, analysis, and web publishing. You'll use additional ArcGIS apps in other chapters.

In this chapter, you will work with a finished map in ArcGIS Pro that has the locations of urgent health-care clinics in Allegheny County, Pennsylvania. These clinics are (1) federally qualified health centers (FQHCs) that provide subsidized health care for underserved populations and (2) nonsubsidized clinics that provide health care (called urgent care clinics in this book). You will become familiar with the software, learn how the map works, and analyze the locations of both types of health-care clinics to see whether they are sited in appropriate locations.

Part 1
Chapter 1
Tutorial 1

Tutorial 1-1: Getting an overview of ArcGIS Pro

Before starting work on your computer, review this terminology for ArcGIS Pro projects and spatial data.

- A *feature class* is the basic building block for displaying geographic features on a map. You can think of a feature class as a homogeneous layer on a map. Feature classes are vector data and have corresponding attributes for each feature. For example, you will work with a point feature class named **FQHC Clinics** that has points for all FQHCs in Allegheny County. In addition, **Streets** is a line feature class that has centerlines for all streets in the county, and **Municipalities** is a polygon feature class that has boundaries for all municipalities in the county.
- A *raster dataset* (or *raster*) is a major type of spatial data. A raster is an image made up of pixels—squares so small that you can't see them until you zoom in. A common example of a raster is satellite imagery. A raster encoded with geographic coordinates can be used as a layer in a map.
- A *file geodatabase* is a folder with the extension .gdb that stores feature classes, raster datasets, and other related files. Although many other file formats are used for storing spatial data, the file geodatabase is a preferred Esri format. The data used in the Tutorial 1-1 project is in the **Chapter1** file geodatabase, stored in the **Chapter1\Tutorials** folder on your computer.
- A *project* is a file with the extension .aprx that contains one or more maps and related items. For example, you'll open **Tutorial1-1.aprx** after this introduction. This project has two maps, **Health Care Clinics** and **Health Care Clinics_3D**, plus other project items. A project doesn't contain spatial data (such as the feature classes you may use on your maps)—it simply pulls in spatial data that's stored elsewhere, such as in a file geodatabase. A project can be stored in a location of your choice.

Set up the Tutorial 1-1 project

You'll start by opening a project in ArcGIS Pro. The opening map has 12 layers available for Allegheny County but only one turned on—a layer for population density by census tract from the 2020 census. The city of Pittsburgh is in the center of the county.

1. Browse to links.esri.com/GISTforPro3.4Data to download the tutorial data for the book. Download and extract the files to your local drive.

2. Start ArcGIS Pro on your computer.

3. Sign in with your ArcGIS account username and password. If you don't already have a license, see the licensing options at the beginning of this book in the section "About This Book."

4. Click **Open another project**. In the **Open Project** pane, browse to \EsriPress\GISTforPro\Chapter1\Tutorials and double-click **Tutorial1-1.aprx**.

   The project opens and displays a map called **Health Care Clinics**.

5. On the **Map** tab, click **Bookmarks** > **Allegheny County**.

   The bookmark centers the map to fill the view.

6. In the **Contents** pane, scroll down to see the legend for the **Population Density** layer.

   This groups the census tracts into nine classes. A white-to-black color scheme shows how many people live there per square mile, in increments of 1,000.

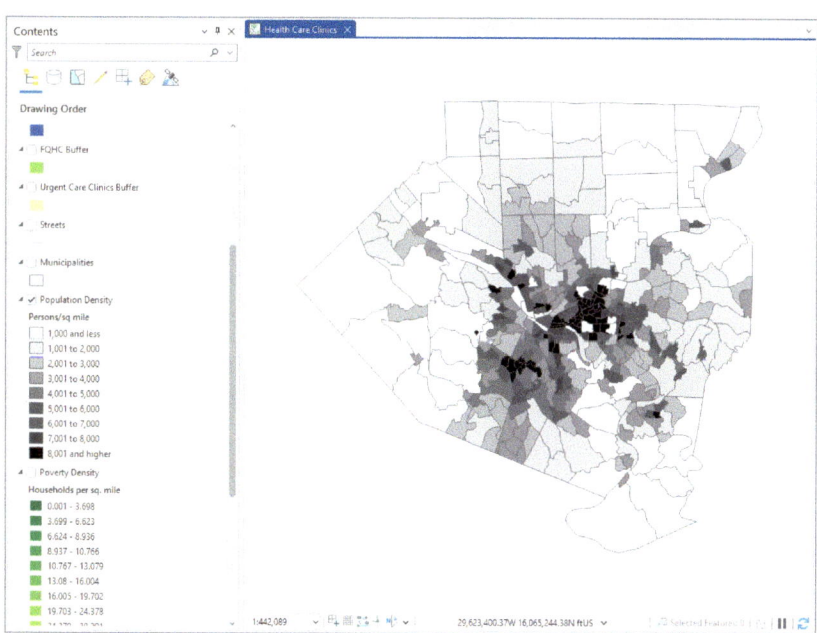

7. On the **Project** tab, click **Save Project As**, browse to **Chapter1\Tutorials**, and save the project as **Tutorial1-1YourName.aprx** (substitute your name for YourName).

   You'll generally open and save projects for each chapter this way. If you make a mistake, you can start again with the original project.

Add and remove a basemap

A basemap is a layer that helps orient users to a location. Map designers place additional feature classes on top of a basemap to provide specific information for visualization, analysis, or problem solving. Although you can create your own basemap, Esri provides the basemaps you'll use in this book. These basemaps are stored in ArcGIS Online.

1. On the **Map** tab, in the **Layer** group, click **Basemap**.

    You'll see a variety of basemaps—available basemaps depend on licensing.

2. Click the **Streets** basemap to add the basemap to your map.

    The **Population Density** feature class covers most of the **Streets** basemap.

    Because the basemap doesn't add useful information in this case, you'll remove it.

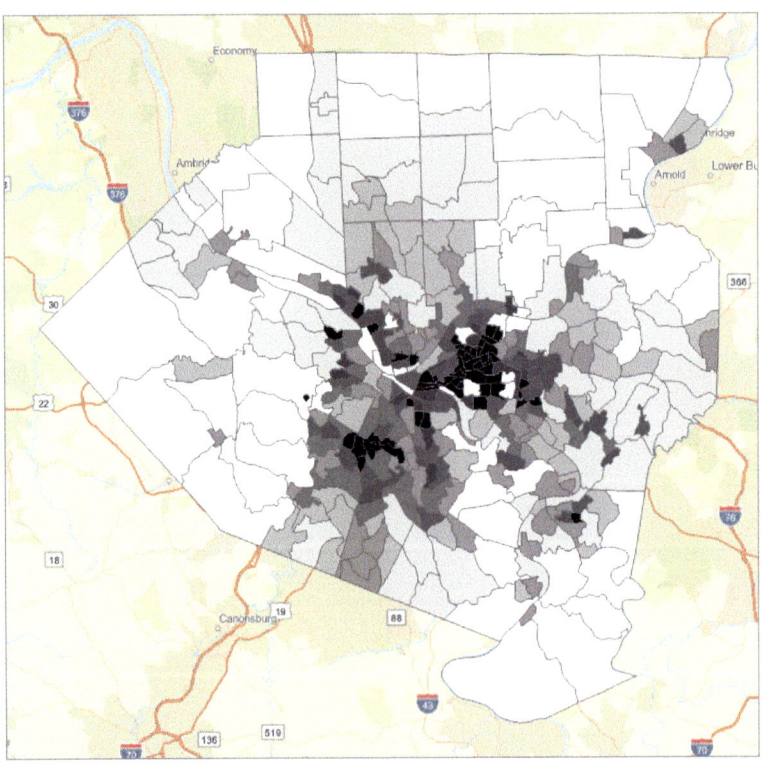

3. Scroll to the bottom of the **Contents** pane, right-click **World Street Map**, and click **Remove**.

> **YOUR TURN**
>
> The Your Turn assignments in this book ask you to repeat the steps just completed but with some modifications. These assignments help you retain the workflows in the steps. Often, you'll need to complete the Your Turn assignments so you can use their results in the next tutorial steps, so don't skip Your Turn assignments.
>
> For this Your Turn assignment, add and remove several basemaps of your choice. You'll notice that some basemaps, such as **Light Gray Canvas**, add a labeling reference layer at the top of the **Contents** pane. When you remove those basemaps, remove the corresponding labeling layer, too.

Turn layers on and off

The order in which feature classes are drawn is shown in the **Contents** pane. The feature class at the bottom of the pane is drawn first, and each layer above it is drawn in turn on top of the preceding layer until you reach the top of the list. Feature classes that cover large areas, such as **Population Density**, must go on the bottom. Other feature classes that may be covered by other layers—such as **FQHC Clinics**, which has points—must go higher in the list to be visible.

1. In the **Contents** pane, scroll down to see the legend for **Population Density**.

   The check mark on the left of **Population Density** indicates that the feature class is turned on.

2. In the **Contents** pane, check the boxes for **Urgent Care Clinics**, **FQHC Clinics**, and **Poverty Risk Area**.

   The three feature classes you just turned on are the subject of this map and show the locations of urgent care clinics relative to poverty risk areas. Right away, you'll notice that the subsidized FQHC clinics tend to be concentrated in areas of high population density (urban) and poverty risk, whereas the non-subsidized urgent care clinics are mostly spread out in areas of low population density (suburbs). Areas inside the poverty risk area polygons have high proportions of poor populations, as determined by the number of households below the poverty income threshold (using average American Community Survey data for the five-year period ending in 2019). The locations of FQHC and urgent care clinics are from 2022.

3. Turn on feature classes that provide the spatial context of where subject features are located: **Pittsburgh**, **Allegheny County**, **Rivers**, and **Streets**.

   Streets, an important spatial context feature class, won't display until the map is zoomed in to a small area (you'll learn about zooming later in this chapter). There are too many detailed streets for viewing at full extent.

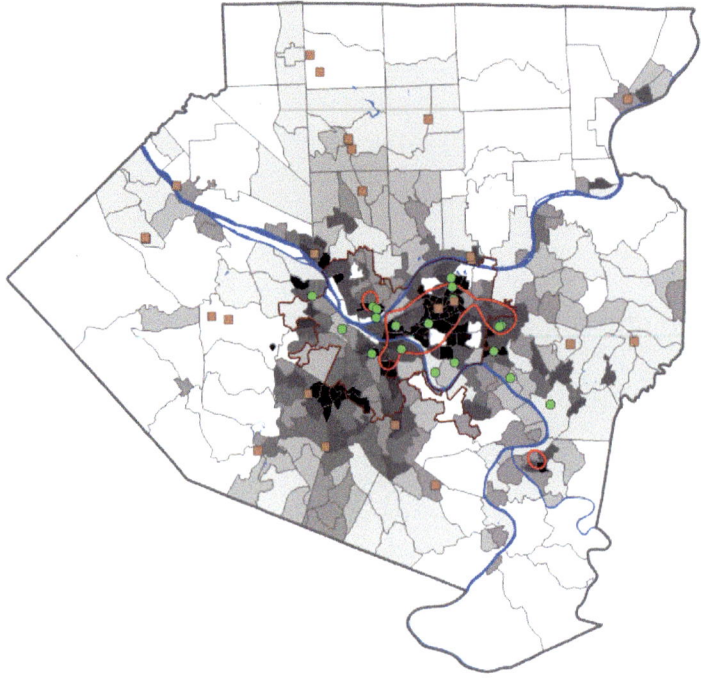

Reorder feature classes in the Contents pane

To make the point that ArcGIS Pro draws from the bottom up in the **Contents** pane, you'll temporarily drag **Population Density** higher to cover other feature classes.

1. Drag **Population Density** to the top of the **Contents** pane, under **Health Care Clinics**.

   Now this feature class covers all other feature classes in the map.

2. Drag **Population Density** back to directly above the **Poverty Density** feature class.

3. At the top of the screen, above the ribbon, click the **Save Project** button.

Some ribbon buttons don't have labels. To identify the button you need, position the pointer above the button to view a pop-up ScreenTip. The **Save Project** button saves the entire project.

Examine the Catalog pane and open and export a map layout

The **Catalog** pane provides access to all components in an ArcGIS Pro project.

1. On the **View** tab, in the **Windows** group, click **Catalog Pane**.

   The **Catalog** pane appears.

2. If the pane isn't docked on the right side of the ArcGIS Pro window, right-click the top of the pane and click **Dock**.

3. In the **Catalog** pane, click the arrows on the left of both the **Maps** and the **Layouts** folders to expand the folders, revealing what's been built so far for this project.

   You are viewing the **Health Care Clinics** map, but you'll also view a 3D version of the same map at the end of this chapter.

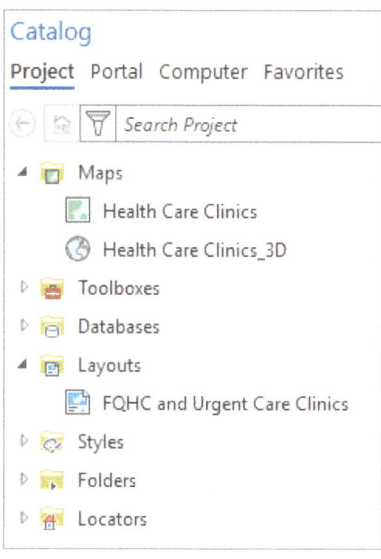

4. In the **Catalog** pane, under **Layouts**, double-click **FQHC and Urgent Care Clinics**.

   ArcGIS Pro displays the layout on a new tab, next to the tab for the map. The map is the main element of a layout, which also includes the title, legend, and scale bar. You'll learn about making layouts later.

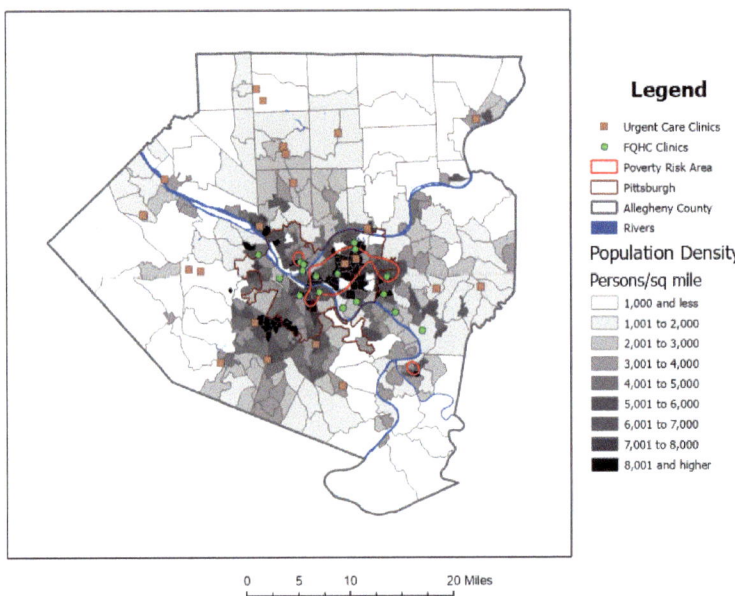

5. At the upper right of the **Catalog** pane, click the **Auto-Hide** button (the pushpin) to temporarily hide the pane.

   The **Auto-Hide** button will collapse the panes along the edge of the screen. You can restore and hide the **Catalog** pane as needed by clicking the **Catalog** tab.

   Next, you'll use a tool to export the map layout as an image file that can be used in a report or presentation or on a website. ArcGIS Pro has hundreds of tools, and you'll use many of them in this book. Each tool has a dialog box (a pane) for values (parameters) you need to add to run the tool, generally by choosing values from lists or entering values by typing.

6. On the **Share** tab, in the **Output** group, click **Export Layout**.

   The **Export Layout** tool pane opens, ready for you to add and type parameters as needed to export an image file of the layout.

7. Complete the following steps to add parameters:
   - On the **File Type** drop-down menu, click **PNG**.
   - For **Name**, click the **Browse** button, browse to save the file to your desktop, and rename it **FQHCAndUrgentCareClinics.png**.
   - For **Resolution**, type **150**.
   - On the **Color Depth** drop-down menu, click **24-bit True Color**.
   - Click **Export** to run the tool.

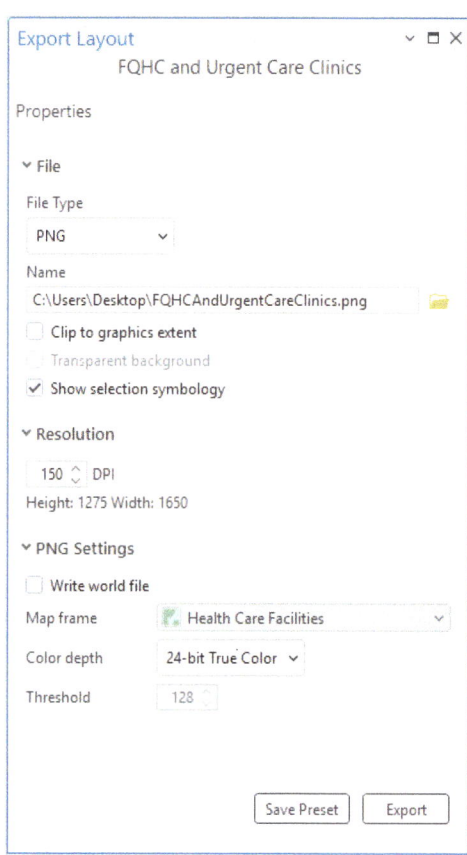

8. On your desktop, double-click **FQHCAndUrgentCareClinics.png** to open the image in a photo viewer.

9. Examine the image and close the photo viewer.

10. Click **Save Project** to save your project.

YOUR TURN

Next to the **FQHC and Urgent Care Clinics** layout tab, click the **Health Care Clinics** map tab to activate it. In the **Contents** pane, turn on the **FQHC Buffer** and **Urgent Care Clinics Buffer** feature classes. The **FQHC Buffer** shows circles with a one-mile radius, and the **Urgent Care Clinics Buffer** shows circles with a five-mile radius, each drawn around its respective clinics. Allegheny County, outside Pittsburgh, consists mostly of suburbs for which expected travel distances are greater than distances in urban areas. The buffers are partly transparent, allowing you to see the population density below them. By comparing the clinic service areas and the population density, you can get an idea of how services are apportioned. Next, switch back to the **Layout** tab. ArcGIS Pro has already added the newly displayed feature classes to the layout map and legend. When you've finished, save your project.

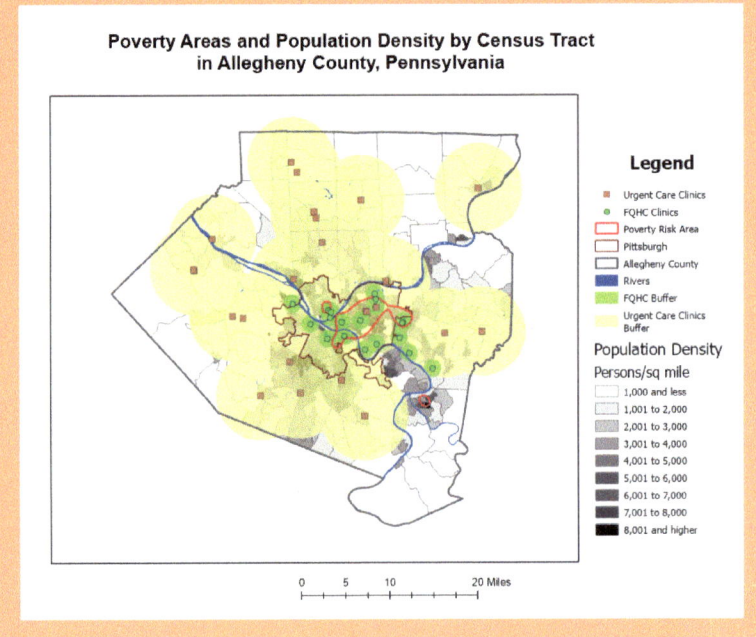

Tutorial 1-2: Navigating ArcGIS Pro

In ArcGIS Pro, you can zoom (go directly) to any part of a map, drag (pan) the map to a different location, and zoom in or out. You can set some feature classes, such as streets, to appear only when you are zoomed to a certain scale, and you can set your scale to turn off other feature classes. You can access preset locations and scales using spatial bookmarks. You can read the attribute data of any feature by clicking the feature to show a pop-up with that information. Last, you can search for features by using attribute values, such as the name of a street.

Set up the Tutorial 1-2 project

1. Above the ribbon, at the top left, click **Open** and browse to **Chapter1\Tutorials**.

2. Open **Tutorial1-2.aprx** and save the project as **Tutorial1-2YourName.aprx** in **Chapter1\Tutorials**.

Use a pop-up window

1. On the **Map** tab, in the **Navigate** group, click **Full Extent** and click the **Explore** button.

2. On the map, click the point representing the urgent care clinic farthest to the north to show a pop-up window with attribute data for that feature. Click the pop-up's website hyperlink and read about that clinic. When you've finished, close your browser.

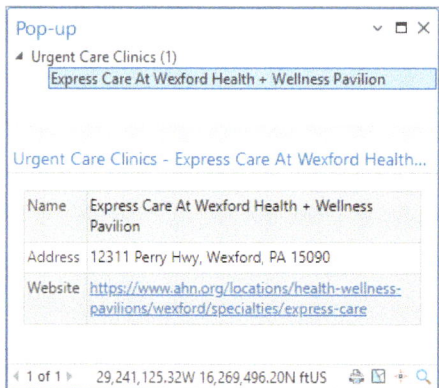

3. Drag the pop-up off your map, to the side.

4. Hover your pointer over each of the buttons in the lower right of the pop-up to read what they do and click the **Zoom to this feature on the active map or scene** button several times.

   The map centers and zooms in on the clinic. If you zoom in close enough, the **Buffers** and **Population Density** feature classes turn off, the **Streets** feature class turns on, and the clinic is labeled. If you zoom in even more, the streets are labeled. These feature classes and labels have visibility ranges that control whether they are visible.

   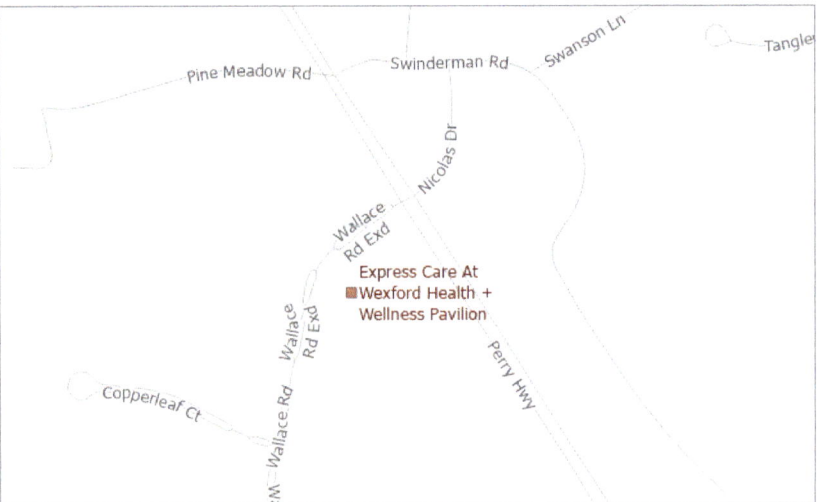

5. Close the pop-up and zoom to the full extent.

Zoom in

1. Position the pointer over the point where the three rivers join in the middle of the map and use the wheel button (or **Plus** key) to zoom in on the map, stopping several times before zooming in more.

   The wheel button (or **Minus** key) is used to zoom out.

2. Scroll up and down in the **Contents** pane.

   Feature classes not drawing at this scale have gray check marks, whereas feature classes that are drawn have black check marks.

3. Click and pan the map to a new location.

Your pointer is automatically in pan mode when the **Explore** button (on the **Map** tab, in the **Navigate** group) is activated. You can also use the arrow keys on your keyboard to pan the map. If you have a touch screen, tap and slide the map to pan.

4. On the **Map** tab, in the **Navigate** group, click the **Previous Extent** button (left-pointing arrow) a couple of times.

   Clicking this button moves you back through the sequence of steps you've taken. There's also a **Next Extent** button (right-pointing arrow) used to advance through your sequence of steps.

5. Zoom to the full extent.

Zoom in to a raster feature class

All the layers, except the **Poverty Density** layer, are vector data, made up of points, lines, or polygons. ArcGIS Pro draws vector data on the fly from stored geometry and drawing instructions, including how the map designer wants them symbolized (such as the brown square symbols for **Urgent Care Clinics**). Rasters, however, are image data (for example, JPEG and TIFF) and are rendered pixel by pixel.

1. Turn the feature classes on or off as needed, so that only the following feature classes are on: **Poverty Risk Area**, **Pittsburgh**, **Allegheny County**, **Rivers**, **Streets**, and **Poverty Density**.

   Tip: Press and hold the **Ctrl** key and click a feature class check box to turn all feature classes on or off. Then adjust for the desired map.

Poverty Density shows households below the poverty income level per square mile. The raster is rectangular, as are all raster datasets. If you look closely, you'll see that the **Poverty Risk Area** feature class (which is drawing on top of the raster) appears to be encircling the deep-red areas of the **Poverty Risk** raster dataset.

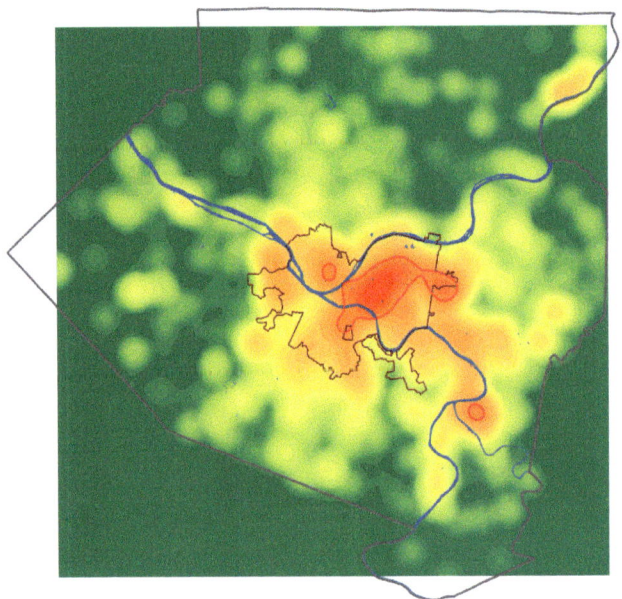

2. Zoom in to the center of the map until you can see the individual pixels of **Poverty Density**.

   In the figure, the pixels of the **Poverty Density** raster map are next to the smooth vector edge of a **Poverty Risk Area** polygon.

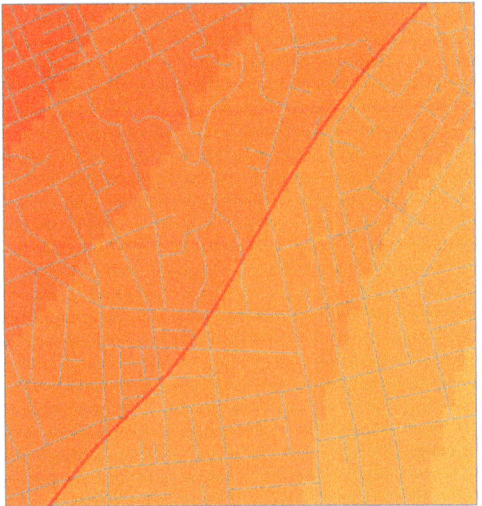

3. Zoom to the full extent and turn on all feature classes except **Municipalities** and **Poverty Density**.

Use bookmarks

Spatial bookmarks allow you to zoom to preset map views.

1. On the **Map** tab, in the **Navigate** group, click the **Bookmarks** button.

   Three bookmarks are available for the open map: **Allegheny County**, **Poverty Areas**, and **Pittsburgh East End**. Also available are two bookmarks for a 3D map that you'll use in tutorial 1-4.

2. Click the **Poverty Areas** bookmark.

   The map zooms to that area. Your map may have different feature classes displayed, depending on the size of your computer's screen and map window, because the thresholds that switch feature class displays on and off depend on the ratio of feature sizes on the screen to actual feature sizes on the ground. You'll learn more about map scale in chapter 2.

3. In the lower right of the map, zoom in and pan to the poverty risk area there until streets appear.

   *Tip: An alternative is to press **Shift** and draw a rectangle around the area desired for closer viewing.*

4. On the **Map** tab, in the **Navigate** group, click **Bookmarks** > **New Bookmark**.

5. In the **Create Bookmark** window, for **Name**, type **McKeesport Poverty Area** and click **OK**.

6. Click the **Allegheny County** bookmark and then try your new bookmark.

7. Click **Bookmarks** > **Manage Bookmarks**.

8. In the **Bookmarks** pane, alphabetize the **Health Care Clinics** bookmarks by dragging to rearrange them.

9. Close the **Bookmarks** pane.

10. Zoom to the full extent.

Search for a feature

Next, you'll use the ArcGIS Pro query builder for Structured Query Language (SQL) queries. SQL is the standard language for querying tabular data. In this quick preview, you'll search for locations based on their attribute data values.

1. In the **Contents** pane, uncheck the **Population Density** check box and check the **Municipalities** check box.

2. Right-click **Municipalities** to open the menu and click **Attribute Table**.

   Every vector feature class has an attribute table, and each feature (point, line, or polygon) of a feature class has a record or row of data.

3. On the **Map** tab, in the **Selection** group, click the **Select By Attributes** button.

   The **Select By Attributes** tool appears. **Municipalities** is already selected as the input. In the box for building the query, the SQL expression starts with the word *Where*.

4. Complete the query as follows:
   - After **Where**, click the down arrow and click **NAME**.
   - Keep **is equal to**.
   - Click the last down arrow and click **McKees Rocks**.

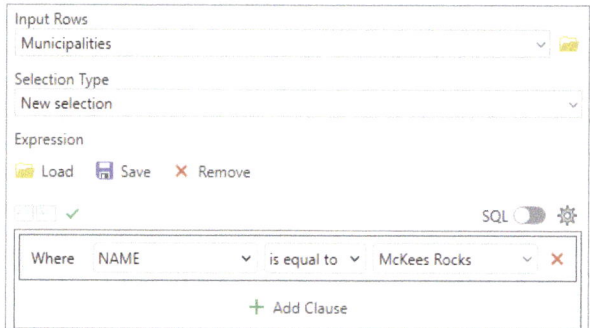

   - Click **OK**.

The result is that the **McKees Rocks** record and feature are selected. You will complete the next two steps to see the record and feature.

5. At the bottom left of the **Municipalities** attribute table, click the **Show Selected Records** button.

Only the selected record is visible now.

6. In the **Contents** pane, right-click **Municipalities**, point to **Selection**, and click **Zoom To Selection**.

7. At the top of the **Municipalities** table, click the **Clear** button and close the table.

8. Uncheck the **Municipalities** check box and check the **Population Density** check box.

> **YOUR TURN**
>
> Search for an FQHC, the **Birmingham Free Clinic**, and zoom to that clinic. In the **Select By Attributes** pane, for **Input Rows**, make sure you have selected **FQHC clinics**. When you finish, clear your selections, close any open panes or tables, and zoom to the full extent. Save your project.

Tutorial 1-3: Working with attribute data

Attributes play a major role in GIS. Besides providing data needed to solve a problem or investigate spatial patterns, attributes allow you to search for useful information and mapped features.

Open a table

1. Open **Tutorial1-3.aprx** and save the project as **Tutorial1-3YourName.aprx**.

2. In the **Contents** pane, right-click **FQHC Clinics** and click **Attribute Table**.

   The table shows 19 FQHC clinics with name and address data, latitude and longitude coordinates, and website addresses. The first five records are shown in the image.

	OBJECTID *	Shape *	Name	Address	Latitude	Longitude	Website
1	1	Point	East End Community H...	745 N Negley Ave, Pitts...	40.471228	-79.926983	https://www.freeclinics.(
2	2	Point	Allies for Health + Wel...	5913 Penn Ave 2nd Flo...	40.462022	-79.926167	<Null>
3	3	Point	Sto-Rox Family Health...	710 Thompson Ave, M...	40.465029	-80.062758	https://www.storox.org/
4	4	Point	Center for Inclusion H...	320 E North Ave, Pittsb...	40.457079	-80.003669	<Null>
5	5	Point	North Side Christian H...	816 Middle St, Pittsbur...	40.454978	-79.999378	https://mycw4.eclinicalw

3. Adjust the column widths in the table if you can't read the full cell contents by positioning the pointer between column names on the top row until the pointer becomes a two-headed arrow and click and adjust by moving the column boundary left or right.

   You can also double-click when you see the two-headed arrow to automatically resize the column widths.

4. In the table, drag the **Website** column and place it before **Latitude**.

5. Right-click the **Name** column heading and click **Sort Ascending**.

6. Close the **FQHC Clinics** table and save the project.

> **YOUR TURN**
>
> Turn off all layers and then turn on **Population Density**. Open the attribute table for **Population Density**. The **GEOID** attribute is a geocode (unique identifier or primary key) assigned to census tracts by the US Census Bureau. Attributes of interest are **Population** (2020 population), **AreaSqMiles**, and **PopDensity**, which is the total population divided by the total area. Using sorting, find the tract with the highest population density, 29,492.7 persons per square mile. Select that record (click the gray square on its left). You can see the corresponding census tract in the selection color (cyan) on the map. Click the Clear button at the top of the table. Close the table when you're finished.

Work with the fields view of an attribute table

You can change the order of attributes (columns) in a table, change the names and displayed names (aliases) of attributes, see the data type of attributes, delete attributes, and make only certain attributes visible to the user—all using the **Fields** view of a table.

1. In the **Contents** pane, turn off **Population Density** and turn on **Municipalities**.

2. Open this feature class's attribute table.

   All attributes are those provided by the Census Bureau. You'll learn more about census data later. The only attributes you need to see and use are **GEOID** and **NAME** (municipality name). Other attributes are needed by ArcGIS Pro for processing—namely, **OBJECTID**, **Shape**, **Shape_Length**, and **Shape_Area**—and the rest can be deleted or hidden from view.

3. In the upper right of the table, click the **Menu** button (three stacked lines) and click **Fields View**.

A new tab opens with the field of **Municipalities** and their properties.

Visible	Read Only	Field Name	Alias	Data Type	Allow NULL	Highlight	Number Format	Domain	Default	Length
✓	✓	OBJECTID	OBJECTID	Object ID			Numeric			
✓	☐	Shape	Shape	Geometry	✓					
✓	☐	STATEFP	STATEFP	Text	✓					2
✓	☐	COUNTYFP	COUNTYFP	Text	✓					3
✓	☐	COUSUBFP	COUSUBFP	Text	✓					5

4. Drag the **MTFCC** row using its gray, left-most box to the row above **STATEFP**.

    That action, when saved, permanently rearranges the attribute table so that **MTFCC** is the first row.

    Next, you'll delete a field from the table. Be careful when deleting fields so that you don't delete fields you may need later.

5. Right-click the box for the **STATEFP** row and click **Delete**.

    ArcGIS Pro strikes through text values in that row to indicate that the field will be deleted when table changes are saved.

6. Turn off visibility for all fields except **GEOID** and **NAME**.

    *Tip: Uncheck the **Visible** box at the top of the first data column and check **GEOID** and **NAME** to make them visible.*

7. For the **NAME** field, under the **Alias** column, type **Municipality Name**.

8. On the ribbon, locate the **Fields** tab. In the **Manage Edits** group, click **Save**, close the **Fields** view, and sort the attribute table in ascending order by municipality name.

	GEOID	Municipality Name ▲
1	4200300724	Aleppo
2	4200303320	Aspinwall
3	4200303608	Avalon
4	4200303928	Baldwin
5	4200303932	Baldwin

9. Close the attribute table.

> **YOUR TURN**
>
> In the **Streets** attribute table, give the **FULLNAME** field the alias **Street Name**, and make only the following fields visible: **FULLNAME, LFROMADD, LTOADD, RFROMADD, RTOADD, ZIPL,** and **ZIPR**. Save your edits and close the **Fields** view and attribute table.

Select records and features of a map feature class

1. Keep **Municipalities** turned on and turn on **FQHC Clinics**.

2. Open the attribute table for **FQHC Clinics** and sort in ascending order by name.

3. In the table, on the left of row 1, click the gray square cell and drag down through row 6 to select the rows.

   You have selected six of the 19 **FQHC Clinics** records in the table and highlighted their point symbols on the map (in cyan), demonstrating the linkage between records and features. Many GIS functions work with selected subsets of records and features.

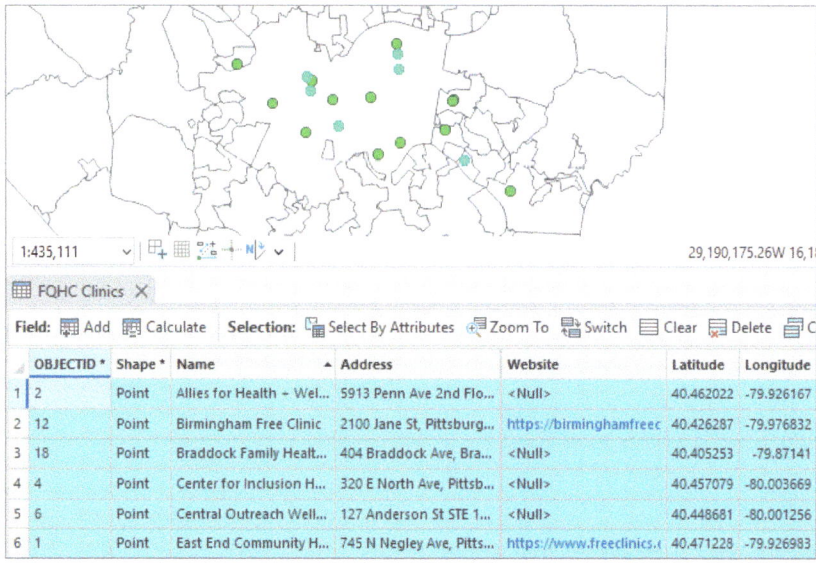

4. At the top of the attribute table, click **Clear**.

5. At the top of the **Contents** pane, click the **List by Selection** button.

6. Uncheck the check boxes so that only **Urgent Care Clinics** and **FQHC Clinics** are turned on.

   Now when you click the map, you can select only clinics and not mistakenly select other features.

7. At the top of the **Contents** pane, click the **List by Drawing Order** button.

8. On the **Map** tab, in the **Selection** group, click **Select**.

9. Press **Shift**, and on the map, individually select any five FQHCs.

   Five corresponding records are selected in the table.

10. At the top of the table, click **Switch**.

    All the records except for the five FQHCs that were originally selected are now selected.

11. Clear the selection.

> **YOUR TURN**
>
> Using the **Selection** tool, draw a rectangle around some FQHCs on the map. All FQHCs within the rectangle are selected. Press and hold **Shift** and draw a different rectangle. More FQHCs are added. Press and hold **Ctrl** and click an already selected FQHC. That FQHC gets unselected, and all other selected FQHCs remain selected. Press and hold **Shift** and reselect the FQHC you just unselected. That FQHC is added back to the selection. Clear the selections and close the table. This assignment showed that you can select any subset of features.

Obtain summary statistics using a tool

ArcGIS Pro has hundreds of tools for processing geographic and related data. You can search for a tool, fill out its form to specify inputs, set parameters that control algorithm behavior, and name and specify where to store outputs. The **Summary Statistics** tool computes common statistics (for example, minimum, maximum, mean, and standard deviation) and writes the results to a new table. Obtaining and studying summary statistics for attributes of interest are among the first steps of any analysis.

1. In the **Contents** pane, turn off **FQHC Clinics** and turn on **Population Density**.

2. On the **Analysis** tab, in the **Geoprocessing** group, click **Tools**.

3. In the **Geoprocessing** pane, click the **Toolboxes** tab.

4. Expand **Analysis Tools > Statistics**.

   Tip: If you know the name of a tool, you can open it by clicking the **Tools** button, and in the **Find Tools** search box in the **Geoprocessing** pane, typing the name or partial name of the tool.

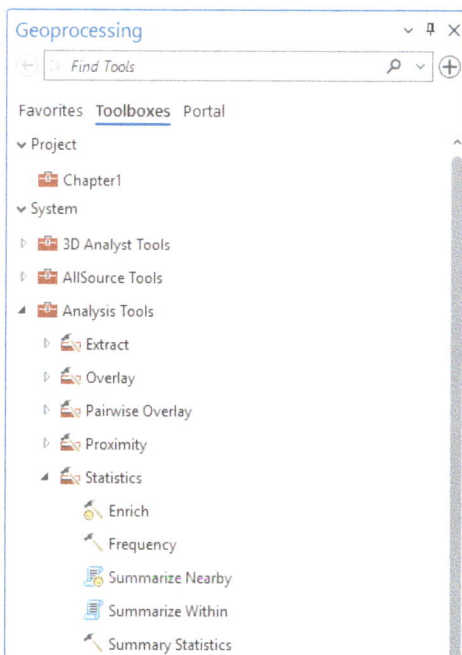

5. Click the **Summary Statistics** tool to open it.

   Once you start applying settings, ArcGIS Pro automatically chooses your project's file geodatabase, **Chapter1.gdb**, and names the output table **PopulationDensity_Statistics**. You can override those defaults if desired. Make sure that **Case Field** is clear.

6. Type or make the selections for the following parameters:
   - For **Input Table**, choose **Population Density**.
   - For **Field**, choose **PopDensity**.
   - For **Statistic Type**, click **Minimum**.
   - Repeat this process three more times, clicking additional **PopDensity** fields. For **Statistic Type**, click **Maximum**, **Mean**, and **Standard Deviation**, respectively.

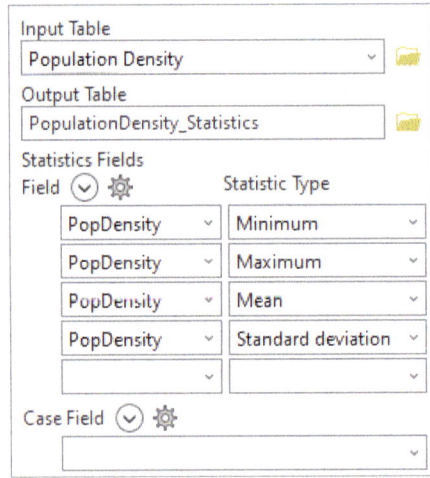

7. Click **Run**.

8. In the **Contents** pane, scroll down, right-click the **PopulationDensity_Statistics table**, and click **Open**.

   At least one tract has no population, so the minimum population is zero, the maximum population density is 29,492.7, the mean population density is 4,516.6 persons per square mile, and the standard deviation is 4,178.

9. Close the table and save your project.

Tutorial 1-4: Symbolizing maps

This tutorial introduces you to the basics of symbolizing maps. You'll change the symbols of feature classes, including the color and size. You'll label features with their name, choose a font and size, and place a halo around labels to improve readability. You'll add a feature class to the map from a geodatabase, symbolize it, and remove it from your map.

Set up the Tutorial 1-4 project

1. Click **Open Project** and browse to **Chapter1\Tutorials**.

2. Open **Tutorial1-4.aprx** and save the project as **Tutorial1-4YourName.aprx** in **Chapter1\Tutorials**.

Symbolize feature classes

1. In the **Contents** pane, right-click **FQHC Clinics** and click **Symbology**.

   The **Symbology** pane appears. The current symbology is **Single Symbol** with a green circle and a dark-gray boundary.

2. In the **Symbology** pane, click the current symbol (the green circle).

   A gallery of symbols opens.

3. Click **Circle 4**.

   The FQHC symbols on the map immediately change to **Circle 4**.

4. At the top of the **Symbology** pane, click the **Properties** tab.

5. Under **Appearance**, change the **Color** to **Leaf Green** (seventh column, fifth row), change the **Size** to **8** pt, and click **Apply**.

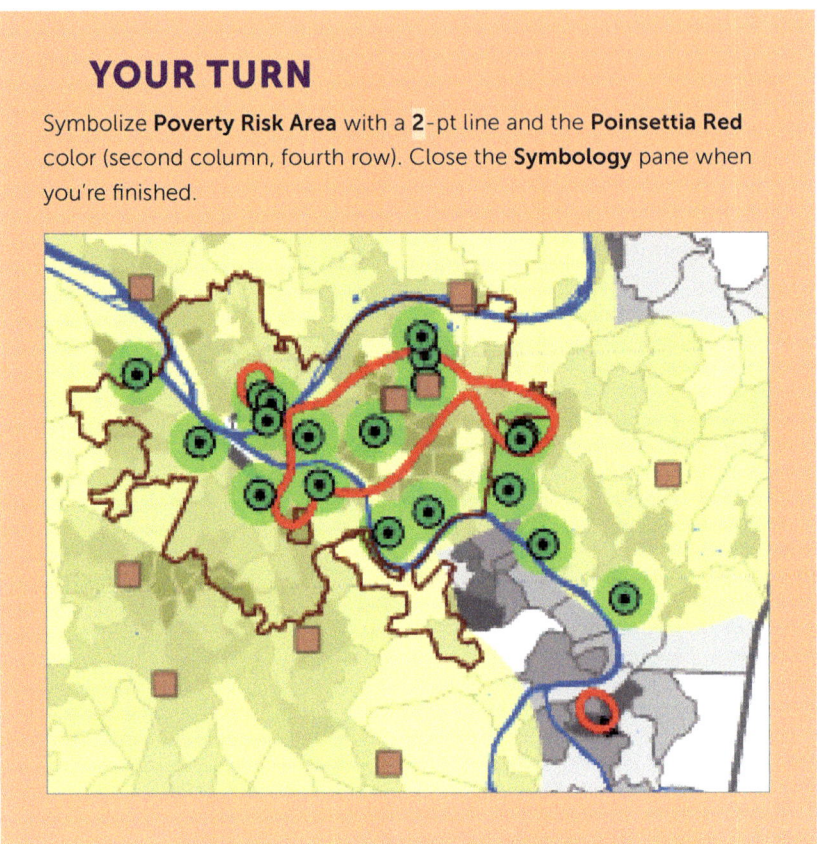

Label a feature class

You will label a feature class using one of its attributes.

1. In the **Contents** pane, turn off **Population Density**, **Urgent Care Clinics Buffer**, and **FQHC Buffer**. Turn on and select **Municipalities**.

    Because you selected **Municipalities** in the **Contents** pane, several contextual tabs appear on the ribbon to give you access to additional functionality for modifying the feature class.

2. Click the **Labeling** tab.

3. On the far left, under the **Layer** group, click **Label** to turn labeling on for **Municipalities**.

    The municipalities are automatically labeled with their **NAME** attribute. Next, you'll make the labels less prominent.

4. In the **Text Symbol** group, change the font size to 7 pt and choose a dark gray.

5. In the **Text Symbol** group, in the lower right, click the **Dialog Launcher** button.

6. In the **Label Class** pane, the **Symbol** tab is active. Scroll down and expand **Halo**. Set the following parameters and then click **Apply**:
   - **Halo symbol**: White fill
   - **Color**: White
   - **Outline color**: No Color
   - **Outline width**: 0.75 pt

7. Zoom to the western part of the county that's centered on the river.

   In effect, the halo erases features surrounding the label so that the label is easy to read.

   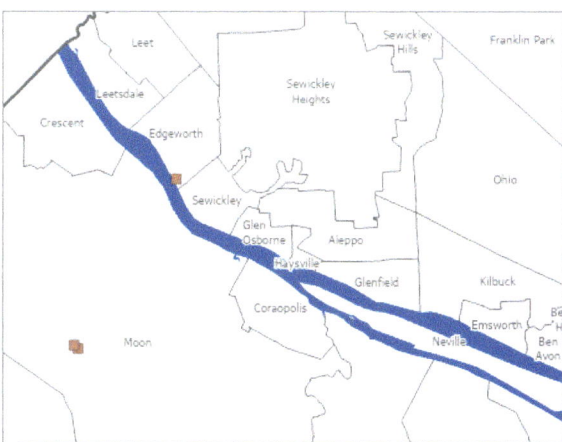

8. Turn **Municipalities** off and zoom to the full extent.

Add and remove feature classes

1. Open the **Catalog** pane.

   *Tip: On the **View** tab, click **Catalog Pane**.*

2. Expand **Databases** and expand **Chapter1.gdb**. Locate the **Parks** feature class and drag it onto the map.

   The **Parks** feature class is added to the **Contents** pane.

3. In the **Contents** pane, reorder **Parks** to be just above **Population Density**.

4. Click the **Parks** symbol (rectangle under **Parks**) to open the **Symbology** pane.

5. In the **Symbology** pane, in the **Gallery** tab, scroll down and click the **Park** symbol (light green with no outline).

6. Zoom to the **Pittsburgh East End** bookmark. Zoom in until streets appear.

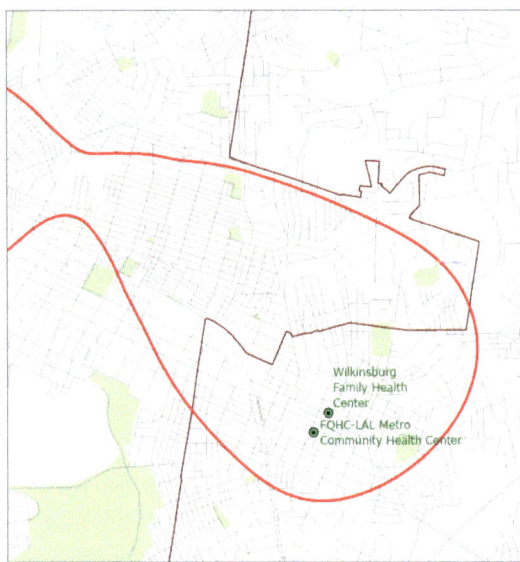

7. In the **Contents** pane, right-click **Parks** and click **Remove**.

8. Zoom to the full extent and save your project.

View and navigate a scene

The Tutorial 1-4 project file has a 3D version of **Population Density** with population density by census tract extruded vertically, which you'll open next. The contrast in population density within an urban area is difficult to appreciate using color symbology. In 3D, the differences are impressive.

1. In the **Catalog** pane, expand **Maps**, double-click **Health Care Clinics_3D**, and wait (several seconds) until the census tracts appear.

   You won't see the 3D effect until you tilt the view in the next step.

2. If you are using a mouse, use the right and left buttons to navigate. If your mouse has a scroll wheel, use that as well. If you are using the keyboard to navigate, press **V** and the down arrow key and tap the down arrow key until the map tilts upward.

The tremendous differences in population density are apparent using the scene.

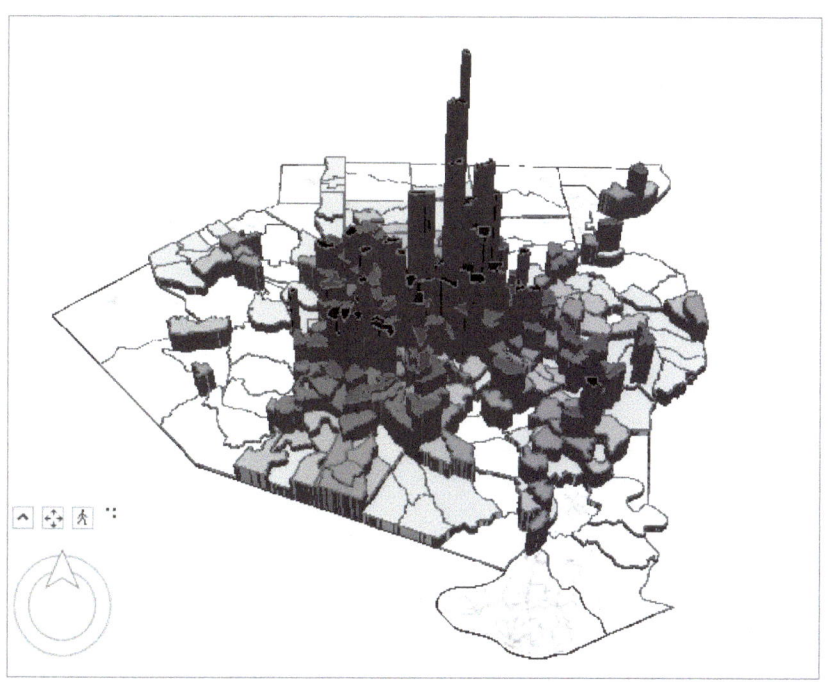

3. Press **V** and the other arrow keys to navigate. Press **V** and the **Plus** or **Minus** key to zoom in or out.

4. Press the **N** key to reorient the map with north up.

5. Press **P** to look straight down at the map.

6. Save and close your project.

Assignments

This chapter has assignments to complete that you can download with data from ArcGIS Online at links.esri.com/GISTforPro3.4Assignments.

CHAPTER 2

Map design

LEARNING GOALS

- Symbolize maps using qualitative attributes and add labels.
- Use definition queries to create a subset of map features.
- Symbolize maps using quantitative attributes.
- Learn about 3D scenes.
- Symbolize maps using graduated and proportional point symbols.
- Create normalized maps with custom scales.
- Create dot density maps.
- Add visibility ranges for interactive map use.

Introduction

In this chapter, you'll learn how to design and symbolize thematic maps. A thematic map strives to solve or investigate a problem, such as analyzing access to urgent care clinics in a region, as you did in chapter 1. A thematic map consists of a subject layer or layers (the theme) placed in spatial context with other layers, such as streets and political boundaries.

Choosing layers for a thematic map requires answering two questions:
1. What layer or layers are needed to represent the subject?
2. What spatial context layers are needed to orient map users to recognize locations and patterns of the subject features?

Often, the subjects of thematic maps are vector layers (points, lines, or polygons), because such layers often have rich quantitative and qualitative attribute data that's essential for analysis. Of course, the subject can be a raster layer (in chapter 10, for example, you'll create a risk index raster map to identify poverty areas of a city, and poverty is the subject of the map). Spatial context layers can be vector, such as streets and political boundaries. These layers can be in either raster or vector formats, including many basemap layers provided by Esri.

The major map design principle for thematic maps is to make the subject prominent while placing spatial context layers in the background. For example, if the subject is a layer with points and you want to focus on them, you could give the point symbols a black boundary and a bright color. These subject features are known as figure and are the main composition of the map. Everything that isn't figure is known as ground. For example, if a context layer has polygons that aren't the focus of the map, you could give the polygons a gray boundary and no color, thus placing them in the background.

Symbology is easy for vector maps because ArcGIS Pro can use attribute values to automate drawing. For example, ArcGIS Pro can draw all food pantry facilities in a city by using unique values with a square point symbol of a certain size and color. Continuing, the software can draw all soup kitchen facilities with a circle of a certain size and different color by using an attribute with type-of-facility code values (including food pantry and soup kitchen).

In this chapter, you will learn to use good cartographic principles on symbology as you build several vector-based thematic maps.

Tutorial 2-1: Symbolizing qualitative attributes

Placing objects of all kinds into meaningful classes or categories is a goal of science. Classifying tabular data is accomplished using attributes with codes that have mutually exclusive and exhaustive qualitative values. For example, a code may have the values low, medium, or high. Any feature with a code is displayed in only one of the classes (the values are mutually exclusive). Moreover, there are no additional size classes (the values are exhaustive). In this tutorial, you'll learn how to symbolize features by their code.

Set up the Tutorial 2-1 project

1. Open **Tutorial2-1.aprx** from **Chapter2\Tutorials** and save the project as **Tutorial2-1YourName.aprx**.

   A **New York City Zoning and Land Use** map opens, showing neighborhoods and a light-gray raster basemap. Two other layers, **ZoningLandUse** and **Water**, are available but not yet visible. None of the layers is properly symbolized yet.

2. Zoom to the **Lower Manhattan** bookmark.

   The subject of this map, zoning and land use, is best viewed and studied at approximately this zoomed scale or closer because the geographic zones are small. You must get close enough to distinguish them from one another.

Display polygons using a single symbol

The **Neighborhoods** and **Water** polygon layers provide spatial context. Context layers are often displayed using outlines with no color fill; water features are an exception and are generally given a blue color and no outline. You will start with **Neighborhoods**.

1. In the **Contents** pane, under **Neighborhoods**, click the white, outlined box to modify the symbol.

2. In the **Format Polygon Symbol** pane, click the **Properties** tab and change **Color** to **No Color**.

3. Change the **Outline color** to **Gray (60%)**, click **Apply**, and close the **Symbology** pane.

> **YOUR TURN**
>
> Turn on the **Water** layer and symbolize the layer with a blue polygon symbol. On the **Gallery** tab, search for **Water** and click one of the **Water (area)** symbols.

Display polygons using unique value symbols

The **ZoningLandUse** layer is the subject of the map. It is symbolized based on the unique values of the primary land-use code. Land-use maps use muted colors, which you'll create next.

1. Turn on **ZoningLandUse**, right-click the layer, and click **Symbology**.

2. In the **Symbology** pane, for **Primary Symbology**, click **Unique Values**.

3. For **Field 1**, choose **LANDUSE2**.

   Random colors are added for six land uses.

   Next, you'll assign colors used by the New York City Planning Department. You'll start by changing the outlines of all polygons from black to light gray. Black outlines often take up too much of your map and distract from the symbolized color. Gray will soften this interference and still show boundaries.

4. In the **Symbology** pane, click **More** > **Format all symbols**.

5. In the **Format Multiple Polygon Symbols** pane, click **Properties**, and change **Outline color** to **Gray (20%)**.

6. Click **Apply** and click the back button to return to the **Symbology** pane.

7. Click the symbol for **Commercial** and change the **Color** to **Rose Quartz** (first row, second column).

8. Click **Apply**, click the back button, and apply the following colors for the remaining land uses:
   - **Manufacturing**: Lepidolite Lilac (first row, 11th column)
   - **Park**: Apple Dust (seventh row, sixth column)
   - **Residential**: Yucca Yellow (first row, fifth column)
   - **Residential/Lt Mfg.**: Soapstone Dust (seventh row, third column)
   - **Waterfront**: Atlantic Blue (ninth row, ninth column)

   Tip: Try right-clicking the color and quickly choosing a new color.

9. Click the down arrow for **More** and click **Include all other values** to clear it.

   All polygons have land-use code values, so this option isn't needed. If left on, other values might be entered in the legend and perhaps confuse map users.

10. Save your project.

You can see all boundaries of primary land uses with their gray outlines.

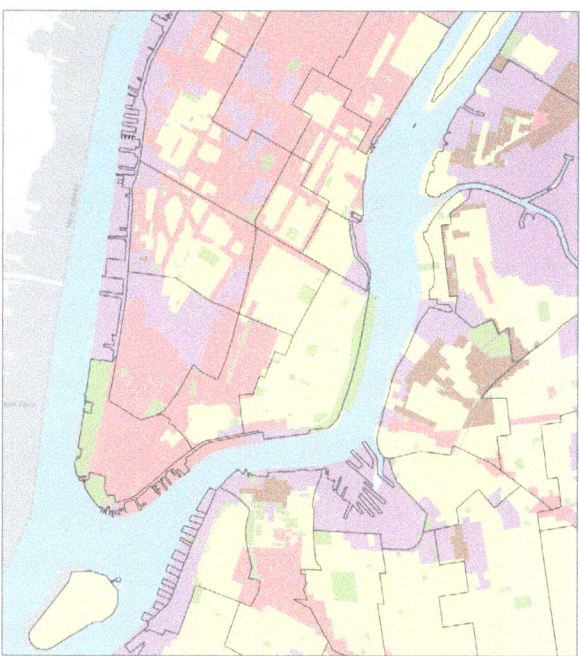

Tutorial 2-2: Labeling features and configuring pop-ups

In addition to graphic elements, such as lines, shaded areas, and symbols, text is also important for conveying information on maps. Two forms of text in ArcGIS Pro are **Labels** and **Pop-ups**. **Labels** are used for identifying graphic elements—for example, providing names of neighborhoods. **Pop-ups** are used for detailed information and include data from several fields, as well as possibly images and charts. In this tutorial, you'll label features and manage pop-ups.

Change label properties

In this section, you'll label all three layers of the map. Each layer has its own label properties and label placements. You'll specify the font, size, color, and placement. **Labels** created from attributes, such as neighborhood names, are an important part of cartography and an integral and informative component of a map.

1. Open **Tutorial2-2.aprx** from **Chapter2\Tutorials** and save the project as **Tutorial2-2YourName.aprx**.

2. Zoom to the **West Village** bookmark.

To maintain visual clarity, labels for a detailed layer such as **ZoningLandUse** are most useful when zoomed in to a neighborhood level or similar larger scale. The labels for these small polygons would overwhelm the map at the full extent of New York City being visualized.

3. In the **Contents** pane, click **ZoningLandUse**.

   The contextual tabs **Feature Layer**, **Labeling**, and **Data** appear on the ribbon.

4. On the **Labeling** tab, in the **Label Class** group, for **Field**, click **ZONE**.

   The **ZONE** field has detailed zoning codes that are familiar to developers and planners.

5. In the **Layer** group, click **Label**. Check the spinning blue indicator at the bottom right of the map and wait for the labels to appear.

   When they appear, the labels are set to default values. You must customize them to better suit the purpose of the map.

6. In the **Text Symbol** group, make the following changes:
   - **Size:** 8 pt
   - **Color:** Gray (50%)

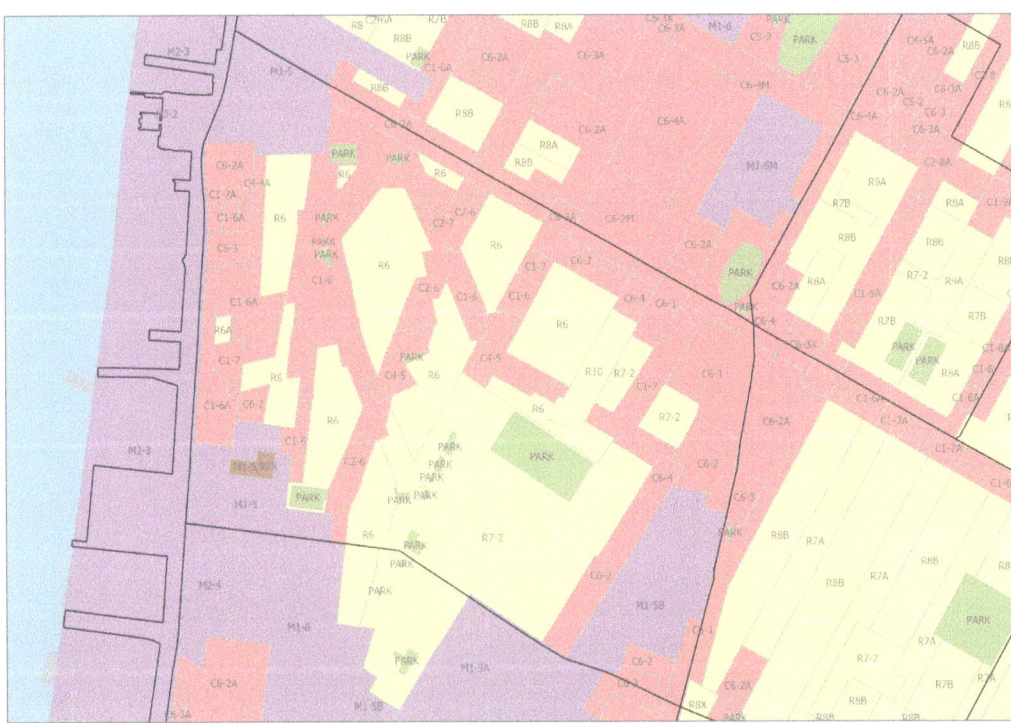

> **YOUR TURN**
>
> Zoom to the **Lower Manhattan** bookmark. You don't want to apply settings as the labels redraw, so apply the following settings first, and then click the **Label** button to turn them on. Label the **Neighborhoods** layer using **Name**, **Arial font**, **Bold**, **11 pt**, and a white halo. Use the **Dialog Launcher** button to open the **Label Class** pane and apply the halo. In the **Label Placement** group, click **Land Parcel**. Label the **Water** layer using **LANDNAME**. Use **Times New Roman**, **Bold Italic**, **12 pt**, and a blue color (we used **Atlantic Blue**: row 9, column 9). This variation of fonts and sizes will make the labels easier to read.

Remove duplicate labels

Labels for some polygons in the **Water** and **ZoningLandUse** layers may overlap with redundant and unnecessary labels. Removing duplicate labels will declutter the map. You will do so in the label properties.

1. In the **Contents** pane, right-click the **Water** layer and click **Labeling Properties**.

2. In the **Label Class** pane, click **Position** (near the top of the pane) and click the **Conflict resolution** button.

3. Expand **Remove duplicate labels** and click **Remove all** to prevent any duplicate labels.

4. For the **ZoningLandUse** layer, set **Remove duplicate labels** to **Remove within fixed distance**. Leave the **Search radius** as the default, **150.0 Points**.

The map now shows only one label per water feature and reduces repeat labels in a fixed distance for the zoning codes.

Manage pop-ups

In this section, you'll turn off pop-ups for layers that aren't the subject of interest (**ZoningLandUse** and **Water**) and configure pop-ups for **Neighborhoods** to display data on population and poverty in neighborhoods.

1. Right-click **ZoningLandUse** and click **Disable Pop-ups**.

2. Do the same for **Water**.

3. Right-click **Neighborhoods** and click Configure Pop-ups.

4. In the **Configure Pop-ups** pane, hover your pointer over **Fields(37)** and click the **Edit pop-up element button** (pencil icon).

   At present, all fields of the **Neighborhood** layer are displayed in pop-ups.

5. Uncheck the **Only use visible fields and Arcade expressions** box and uncheck **Display**.

Now no fields are in pop-ups, so you can select only those you want included.

6. Check the box for the following fields:
   - **BoroName {BoroName}**
   - **Pop {Pop}**
   - **PopPov {PopPov}**
   - **Pop18Under {Pop18Under}**
   - **PopPov18Under {PopPov18Under}**

   You can click and drag fields to reorder them in the resulting pop-up.

7. Close the **Configure Pop-ups** pane.

8. Click anywhere in the **West Village** neighborhood to see its pop-up.

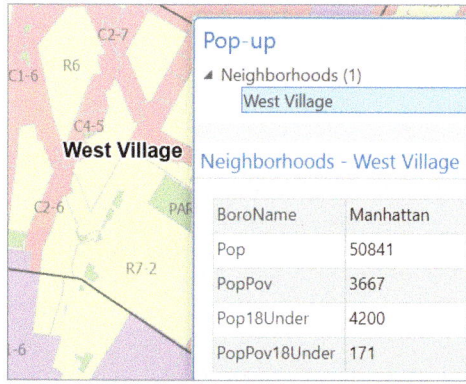

9. Close the pop-up and save your project.

Tutorial 2-3: Filtering with definition queries

Often, a layer has more features than you want to display. If so, you can use a definition query to limit the features to a desired subset of the larger collection, based on values in the feature attribute table. For example, the point features in this tutorial are for all facilities in New York City (food, health care, fire and police, schools, senior centers, and so on). Defining the query allows you to select and display just the features you're interested in. A **Definition Query** is different from **Select By Attributes** in chapter 1. The definition query is used to filter the features of a layer rather than select a temporary subset of features to work with, even though they both use a similar SQL interface.

Create a definition query

In this section, you'll create a definition query to display only a subset of the more than 20,000 government and nonprofit facilities in New York City. **Facilities** is a point layer of locations for services the city provides. The map needs only three types of facilities to be displayed, and you'll select these based on their attribute values. The attribute table contains a field, **Facility_T**, and a series of codes that identify what the facility is used for—for example, 4901 = **Soup Kitchen**, 4902 = **Food Pantry**, and 4903 = **Joint Soup Kitchen and Food Pantry**. Showing the locations of these facilities can help the directors of New York City's food banks determine whether they are located close enough to areas of need in the city.

1. Open **Tutorial2-3.aprx** from **Chapter2\Tutorials** and save the project as **Tutorial2-3YourName.aprx**.

   A **NYC Food Pantries and Soup Kitchens** map opens, showing boroughs, other spatial context layers, and the facilities operated by the city government.

2. In the **Contents** pane, right-click **Facilities** > **Properties**.

3. In the **Layer Properties** window, click **Definition Query** > **New Definition Query**.

4. For **Where**, choose **FACILITY_T**. For the logical operator, choose **is equal to**. For the final value, type **4901**.

   Currently, the definition query contains a single logical condition. Only records that satisfy this condition will appear.

5. Click **Add Clause**. On the new line, choose **Or** as the logical operator, **FACILITY_T** as the field, **is equal to**, and **4902**.

   Now you have two single conditions connected with an **Or** connector to make a compound condition. Any record satisfying one of these two simple conditions will be displayed. If you used the **And** connector here, no records would be selected. A facility cannot have both code values 4901 and 4902.

6. Repeat the process to create another clause: **Or FACILITY_T is equal to 4903**.

   The final compound condition displays a subset of the original table of facilities, with facilities that have just one of the three included values. This example extends to any definition query for selecting subsets of a finite collection of objects classified by a code, such as **FACILITY_T**. If you click the **SQL** button, you can see the SQL code that the definition query generates: `FACILITY_T = 4901 Or FACILITY_T = 4902 Or FACILITY_T = 4903`.

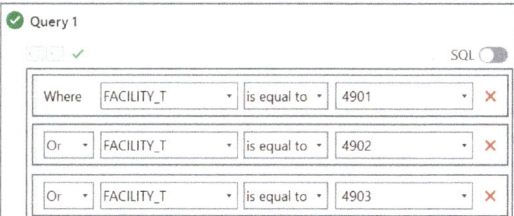

7. Click **OK**.

   The resulting map is a subset (631) of the original 20,000-plus facilities showing just food pantries, soup kitchens, and joint soup kitchens and food pantries. You can verify this by opening the attribute table for **Facilities**.

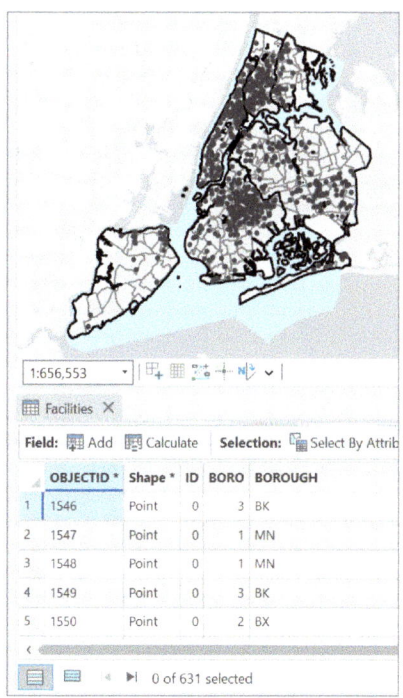

Symbolize figure and ground features

The subject of the next map, **Facilities**, is figure, and all other layers are ground. Figure features are highlighted with bright colors, and ground is in shades of gray.

1. Zoom to the **Manhattan** bookmark.

Someone using this map wouldn't study the map with so many point features at full extent but would zoom in, as shown with this bookmark.

2. In the **Contents** pane, click **Facilities** and rename the layer **Food Facilities**.

3. Right-click **Food Facilities** > **Symbology**.

4. In the **Symbology** pane, set the following parameters:
   - **Primary Symbology**: Unique Values
   - **Field 1**: Factype__1

   This field provides descriptions for the facility codes.

5. Click the **Soup Kitchen** value and drag it to the top.

   Food Pantry should now be in the middle, and **Joint Soup Kitchen** should be at the bottom. The legend in the **Contents** pane reflects that order.

   Next, you'll symbolize the three types of facilities. Using shape and color for point symbols allows for a range of use cases. Users with color blindness can use the shapes to identify facility classes, and the different facilities will remain distinguishable in black-and-white photocopies. Users who can see color get the full effect of shape and color for identifying patterns.

6. Click each symbol, and use the **Symbology Gallery** and **Properties** to change these symbols:
   - **Soup Kitchen**: Square 3, red color, size **8** pt
   - **Food Pantry**: Circle 3, blue color, size **8** pt
   - **Joint Soup Kitchen and Food Pantry**: Cross 3, yellow color, size **10** pt.

7. In the **Symbology** pane, click **More** and turn off **Include all other values**.

   Now the food facilities are sharply in figure with contrasting bright colors and shapes. For example, you can see that Manhattan has more soup kitchens than the neighboring boroughs and that the northern part of Manhattan and the adjoining Bronx have a large cluster of food pantries.

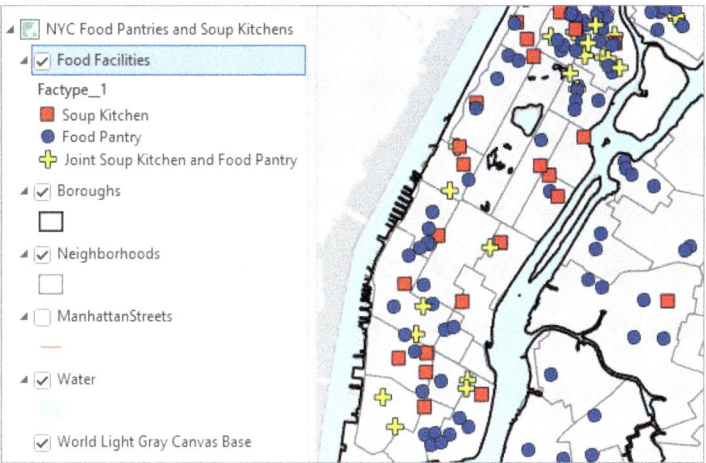

> **YOUR TURN**
>
> A policy decision-maker wants to know the streets where food facilities are located and doesn't want the details of a basemap. Turn off the **World Light Gray Canvas Base** basemap and turn on **ManhattanStreets**. Display **ManhattanStreets** as a ground feature using **Gray (20%)** with a **Line width** of **0.5** pt. Zoom to a few blocks in Manhattan and experiment with various label properties. Save your project.

Tutorial 2-4: Creating choropleth maps for quantitative attributes

The human eye cannot make distinctions unless the changes in graphic elements are relatively large. You must break a numeric attribute into relatively few classes (roughly three to nine), similar to how you create a bar chart for a numeric attribute. Each class has minimum and maximum attribute values. The minimum value is included in the class, but the maximum goes in the next classification. To symbolize map features, you need only the set of maximum values for classes, called break points.

Create a choropleth map of households receiving food stamps

A choropleth map uses color in polygons to represent numeric attribute values. Generally, increasing color value (darkness of a color) in a color scheme represents increasing (higher) values. In this section, you'll use US Census Bureau data of New York City neighborhoods to create choropleth maps for households with persons over age 60 receiving food stamps or Supplemental Nutrition Assistance Program (SNAP) benefits.

Choropleth maps use classification methods to display the data, and methods vary depending on the data and intent of the map. The default classification method is **Natural Breaks (Jenks)**. This method uses an algorithm to cluster values of the numeric attribute into groups, with the boundaries of the groups (break points) defining classes. The Natural Breaks method may be suited for some applications in the natural sciences. However, **Quantile** classification is often a better starting point because the method is easily understood and provides information about the shape of a distribution. The Quantile method breaks a distribution into classes—each with the same percentage of data points. For example, each quartile (quantiles with four classes) has 25 percent of the data observations, with the middle break point being the median.

By studying quantile break points, you can determine whether a distribution is roughly uniform (has equally spaced quantiles) or is skewed to the right (has intervals defined by break points that become progressively larger with larger values). The former become good candidates for the Defined Interval method (uniform distribution with easily read numbers for break points) and the latter for the Geometric Interval method (for an increasing-width interval distribution of break points). Many attributes have skewed distributions.

1. Open **Tutorial2-4.aprx** from **Chapter2\Tutorials** and save the project as **Tutorial2-4YourName.aprx**.

   An **NYC Food Stamps/SNAP Households by Neighborhood** map opens, showing boroughs, neighborhoods, and water features.

2. In the **Contents** pane, click **Neighborhoods** and rename the layer **Over age 60 receiving food stamps**.

3. In the **Symbology** pane for the layer, apply these settings:
   - **Primary Symbology**: Graduated Colors
   - **Field**: O60_FOOD
   - **Method**: Quantile
   - **Classes**: 5
   - **Color Scheme**: Green-Blue (5 classes)

   Tip: In the **Color Scheme** list, check the **Show Names** box.

4. Click the **Histogram** tab.

   Neighborhoods with the highest quintile (quantile with five classes) tend to cluster. The interval sizes generally increase in this case, so a geometric method is an alternative to the Quantile method. You will test the Geometric Interval method in the Your Turn section.

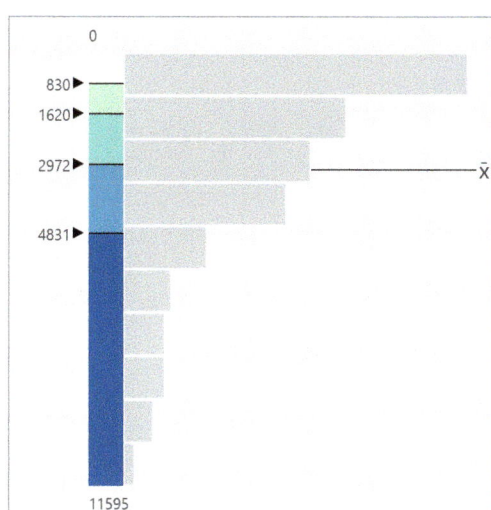

The map clearly shows neighborhoods with a high number of households with persons over age 60 who receive food stamps or SNAP assistance.

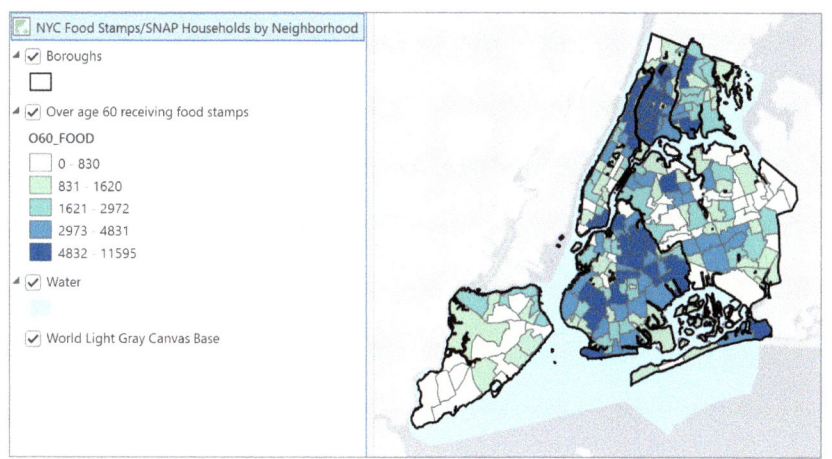

> ### YOUR TURN
>
> Change the symbology **Method** for the choropleth map from **Quantile** to **Geometric Interval**. The break points become larger to provide more detail for the long tail of the distribution. The break points for the Quantile method were **830**, **1620**, **2972**, **4831**, and **11595**. Those for the geometric method are approximately **881**, **2201**, **4181**, **7148**, and **11595**. Select the **Grays (5 Classes)** color scheme.
>
> Try the **Defined Interval** method with an **Interval size** of **2500**. Although perhaps not the best method for this skewed data, this uniform distribution is easier to read, with multiples of 2500 at equal intervals.
>
> Change the **Method** back to **Quantile**.

Extrude a 3D choropleth map

You will learn more about 3D data, scenes, and 3D map navigation later in the book, but now you'll learn to convert a 2D choropleth map into a scene to better visualize data. With 3D, map users can better appreciate extreme values relative to other values, which are not readily apparent by looking at only the color shading of choropleth maps. In 3D, features and layers are often physical features, such as buildings, trees, topography, and so on, but you can display any numeric data as 3D. In this section, you'll convert a 2D map into a scene and display neighborhood polygon features in 3D.

1. On the **View** tab, in the **View** group, click **Convert** > **To Local Scene**.

   A 3D map opens with categories for 3D and 2D layers.

2. In the **Contents** pane, drag **Over Age 60 Receiving Food Stamps** above the **3D Layers** heading.

   You'll begin extruding neighborhood polygons to 3D features using the numeric attribute **O60_FOOD** as the extrusion height.

3. On the **Feature Layer** tab, in the **Extrusion** group, click **Type**, and click **Base Height**.

   Clicking **Base Height** sets the base of the extruded polygons to zero. Next, you'll set the extrusion height to the **O60_FOOD** attribute.

4. In the **Extrusion** group, for **Field**, click **O60_FOOD**.

   Your features and food stamp recipient data are now displayed in 3D.

   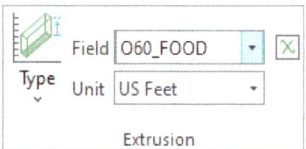

5. Scroll the wheel button on the mouse to tilt the view.

6. Zoom to better see the extruded neighborhoods.

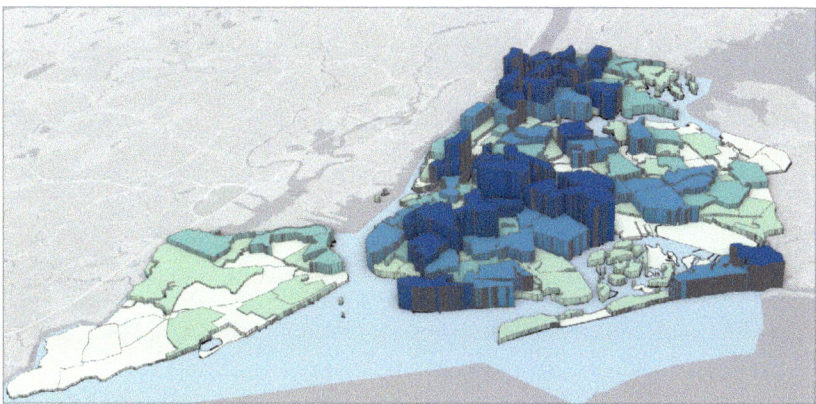

YOUR TURN
Zoom by scrolling the mouse wheel button and pan by dragging the map. Save your project.

Tutorial 2-5: Displaying data using graduated and proportional point symbols

Using ArcGIS Pro, you can display polygon data using point symbols in the center (centroid) of each polygon. In this section, you'll create a map showing the number of food pantries and soup kitchens in New York City neighborhoods as graduated point symbols. The larger the symbol, the more food resources in each neighborhood. You'll also display data as proportional symbols, which are similar to graduated symbols but represent values as unclassified symbols whose size is based on a specific value.

Create a map of graduated-size points

The map in this tutorial uses multiple copies of the **Neighborhoods** feature class. This technique helps you show different attributes of a layer in the same map.

1. Open **Tutorial2-5.aprx** from **Chapter2\Tutorials** and save the project as **Tutorial2-5YourName.aprx**.

   The map shows New York City neighborhoods with persons over age 60 receiving food stamps classified using quantiles.

2. In the **Contents** pane, rename the first **Neighborhoods** layer to **Number of food banks/soup kitchens**.

3. In the **Symbology** pane, apply these settings:
   - **Symbology**: Graduated Symbols
   - **Field**: FOOD_FACIL
   - **Method**: Quantile
   - **Classes**: 5

   The interval width is uniform, 2, except for the last class, so this attribute may be a good choice for the Defined Interval method (uniform distribution). An interval width of 5 is a good choice to include the maximum value of 25. Equal-width intervals are the easiest to read and are, of course, best suited for uniform distributions, but this distribution is not essential. Another possible method is Defined Interval, which allows you to use easily read numbers such as 1, 2, or 5 times 10 to a power (for example, 0.1, 1.0, and 10).

4. Change **Method** to **Defined Interval**.

5. Click the **Template** symbol. Choose **Circle 3** and change the **Color** to yellow.

6. Click **Apply** and click the back button.

   The finished map shows the number of food resources compared with the number of households with persons over age 60 receiving food stamps.

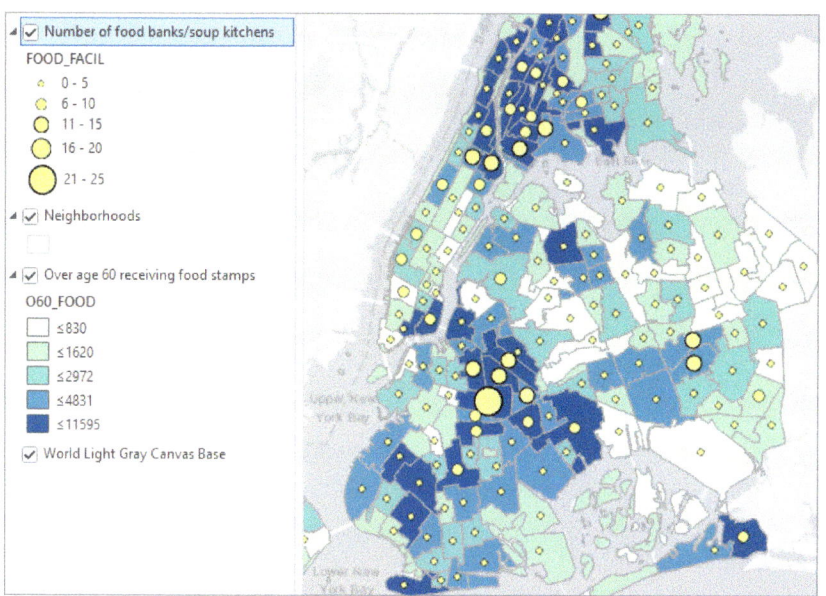

YOUR TURN

Rename the second **Neighborhoods** layer as **Under 18 receiving food stamps**. Change the **Method** to **Proportional Symbols**, **Field** to **U18_FOOD**, **Color** to purple, **Minimum Size** to **2**, and **Maximum Size** to **20**. Use the **Bronx** and **Brooklyn** bookmarks to study the relationship of food banks and soup kitchens and persons over 60 and under 18 receiving food stamps in one map. Save your project.

Tutorial 2-6: Normalizing population maps with custom scales

A choropleth map showing population, such as the number of persons receiving food stamps, is useful for studying needs, such as the demand for goods and services. For example, delivery of food services for the poor requires capacities to match populations, including budgets, facilities, materials, and labor.

Choropleth maps of normalized population data have different uses from those of choropleth maps of populations. Dividing (normalizing) a segment of the population by the total population provides information about the makeup of areas. For example, areas with high proportions of total population receiving food stamps may be better candidates for food pantries and soup kitchens than those with low proportions, because the high-proportion areas are probably poor in many ways, including having poor geographic access to grocery stores and urgent health care.

Other normalized maps could be density maps that divide population and other variables by their polygon areas, yielding a measure of spatial concentration. If you divide a population by its polygon areas, the resulting population density (for example, persons per square mile) can provide information related to congestion or how people are distributed across an area.

A neighborhood with a high density of households receiving food stamps but a low density of food banks and soup kitchens may help determine potential locations for new food banks or kitchens.

In this tutorial, you'll normalize the number of female-headed households (single mothers) with children under the age of 18 receiving food stamps by the total number of households in each neighborhood. You'll find the same information for male-headed households (single fathers) with children under the age of 18 receiving food stamps and compare the two populations using a custom scale.

Create a choropleth map with normalized population and custom scale

You'll create a custom classification, which is often easier to read than other classifications. The Geometric Interval method works well for representing the long tails of distributions skewed to the right, but the break points of this method don't follow a pattern that can be read easily. The custom classification in this tutorial has intervals that double in width (and thus form a geometric progression), which is read easily.

1. Open **Tutorial2-6.aprx** from **Chapter2\Tutorials** and save the project as **Tutorial2-6YourName.aprx**.

   The map shows New York City neighborhoods (female- and male-headed households receiving food stamps) and boroughs.

2. Symbolize **Female headed households receiving food stamps** using these settings:
   - **Symbology**: Graduated Colors
   - **Field**: U18FHHFOOD
   - **Normalization**: TOT_HH
   - **Method**: Quantile
   - **Classes**: 5
   - **Color Scheme**: Grays (5 classes)

   These settings show the fraction of single mothers with children under 18 receiving food stamps. Next, you'll show the values as a percentage.

3. Click the **Advanced symbology options** button. Expand **Format Labels** and apply these settings:
   - **Category**: Percentage
   - **Percentage**: Number represents a fraction
   - **Rounding**, **Decimal places**: 0

4. Click the **Primary symbology** button and click the **Histogram** tab.

   The classification shows the percentage of single mothers who receive food stamps. Expressed as percentages and rounded, the quantile break values are at approximately 1 percent, 2 percent, 6 percent, 10 percent, and 26 percent. Next, you'll create custom classes using the mathematical progression 2 percent, 4 percent, 8 percent, 16 percent, and 26 percent, and the last value is the maximum.

5. Change **Method** to **Manual Interval**.

6. Click the **Classes** tab.

7. In the **Classes** pane, double-click the cell for the first **Upper value**, type **0.02**, and press **Enter**.

   This makes the first class 2 percent.

8. Continue selecting the **Upper values**, and type **0.04**, **0.08**, **0.16**, and **0.26** (the maximum value rounded to two decimal places).

9. Click the **Histogram** tab.

This set of break points has wider intervals for low values than quantiles, thus providing more information about high values.

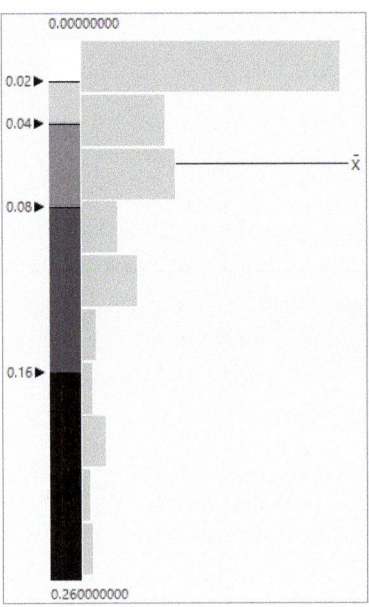

The numeric sequence of the custom break points has both increasing interval widths, as desired for the long-tailed distribution, and a recognized and easy set of values to read.

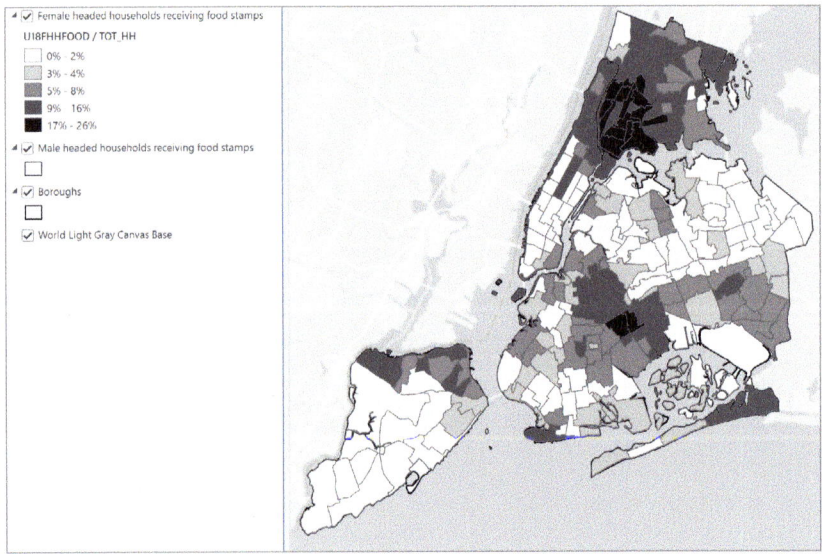

Import symbology and use swipe to compare features

To compare two maps, especially normalized segments of the same total population, you'll need the same symbology in both maps. Next, you'll import and reuse the symbology of the female-headed households for the male-headed households. You'll see fewer male-headed households receiving food stamps.

1. Open the **Symbology** pane for **Male headed households receiving food stamps**, click the **Options** button (three stacked lines), and click **Import symbology**.

2. In the **Apply Symbology From Layer** pane, apply these settings:
   - **Symbology Layer**: Female headed households receiving food stamps.
   - **Target Field**: Change U18FHHFOOD to U18MHHFOOD.

3. Click **Run**.

4. In the **Contents** pane, click **Female headed households receiving food stamps**.

5. On the **Feature Layer** tab, in the **Compare** group, click **Swipe**, and drag the pointer vertically or horizontally to reveal the layer underneath.

   This step allows you to see the values of the **Male headed households receiving food stamps** layer without having to turn off the layer above it. You can see the same spatial patterns in both maps but with lower percentages for male-headed households receiving food stamps.

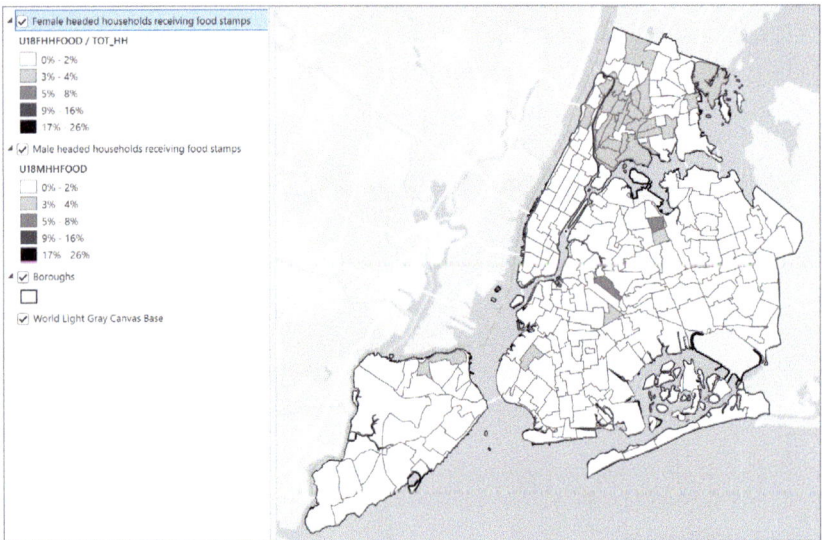

6. On the **Map** tab, click **Explore** to deactivate the **Swipe** tool, and save your project.

Tutorial 2-7: Creating dot density maps

Dot density maps are another way to denote quantitative values. These thematic maps typically display a total number randomly across a statistical unit. An advantage over choropleth maps is that more than one variable can be displayed at the same time using different colored dots. When you use multiple variables, hues should be distinct. Lightness and value should be similar so the map doesn't emphasize one variable over the other.

Neighborhoods with a high number of people receiving food assistance may help determine potential locations for new food banks or soup kitchens. A dot density map can display two populations receiving food stamps (under 18 and over 65) on the same map.

Create the dot density map

1. Open **Tutorial2-7.aprx** from **Chapter2\Tutorials** and save the project as **Tutorial2-7YourName.aprx**.

The map opens with layers added but not yet classified.

2. Symbolize **Persons receiving food stamps** using these settings:
   - **Primary Symbology**: Dot Density
   - **Fields**: U18_FOOD and O60_FOOD
   - **18_FOOD Symbol**: Circle 1, blue, no outline. **Label: Population under 18**
   - **O60_FOOD Symbol**: Circle 1, red, no outline. **Label: Population over 60**
   - **Dot Size**: 2 pt
   - **Dot Value**: 50

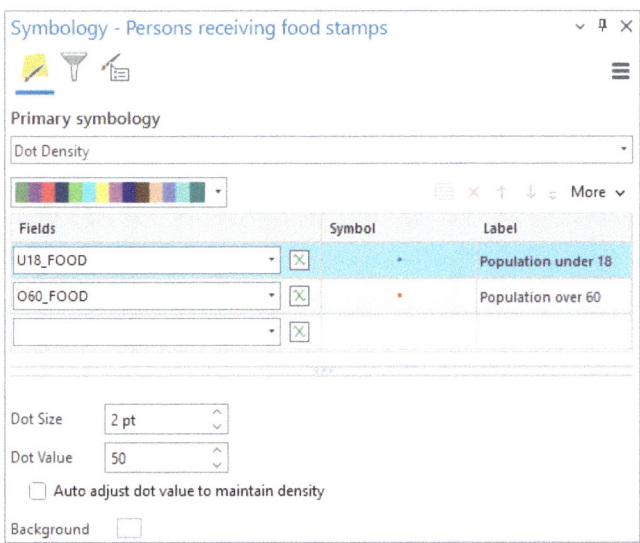

Densities of populations in each age range appear. The dots will resize depending on the scale of your map.

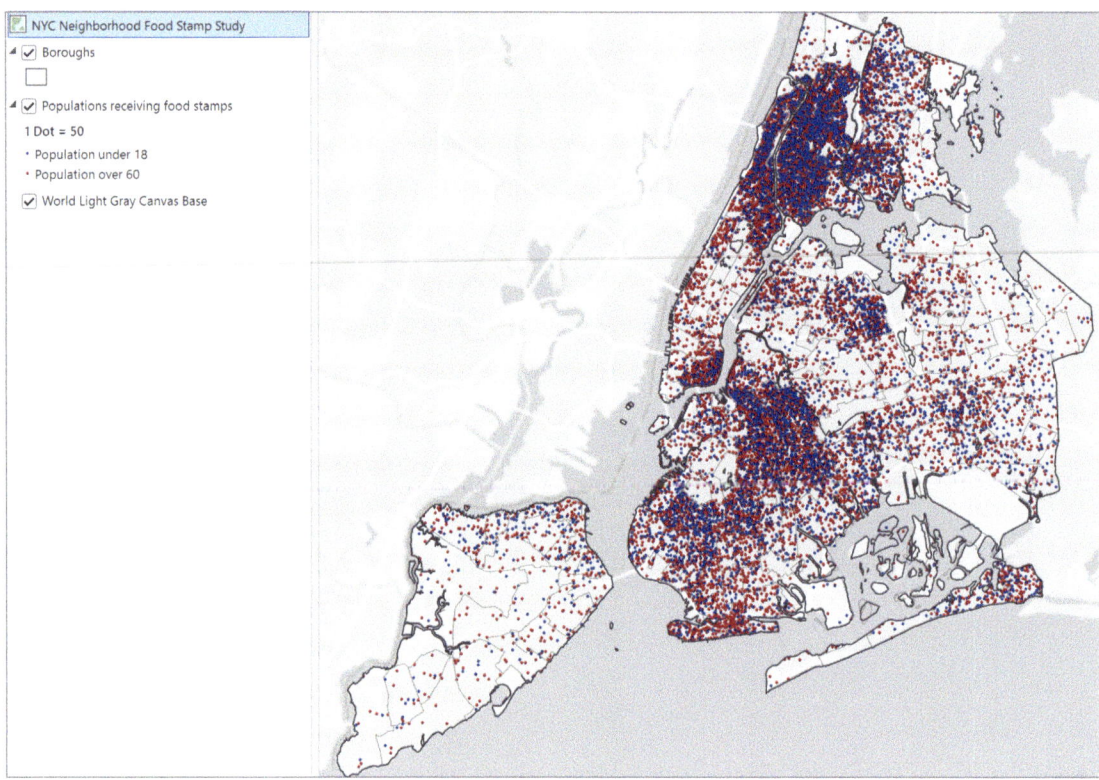

YOUR TURN

Change the **Dot Value** (number of persons that each dot represents) from 50 to **100**. Zoom and pan the map to see the change in density. Save your project.

Tutorial 2-8: Setting visibility ranges

In this tutorial, you'll set visibility ranges for feature labels and map layers. GIS uses visibility ranges to automatically turn layers and labeling on and off, depending on how far you zoom in or out. Before you work with visibility ranges, you'll learn about map scale, the underlying measure that controls visibility ranges.

Suppose that you have two points on your map, A and B, which are one inch apart on your computer screen. Map scale is the ratio of the distance between A and B (one inch) on your computer screen divided by the distance between the same two points in inches on the ground. An example scale for displaying the Lower 48 states is 1:50,000,000, meaning that one inch on your computer's map corresponds to 50,000,000 inches on the ground. Map scale is unitless, as a ratio that divides units. For example, 1:50,000,000 can also mean one foot on your map equals 50,000,000 feet on the ground. You can use any distance unit. The map terms *small scale* and *large scale* are counterintuitive, so understanding the next four sentences is important. The scale 1:50,000,000 is called small scale because the ratio 1/50,000,000 is small, even though the map shows a large area. And the scale 1:24,000 is considered large scale because its ratio is relatively large, even though the map shows a small area. A large scale of 1:24,000 would be useful for showing neighborhoods in a city. All map scales are considerably less than 1 because 1:1 is life size, and all maps represent land areas much larger than the maps themselves.

Next, you'll set visibility ranges to view neighborhood details at a large-scale level so when you zoom in close enough, features and labels will turn on at the metropolitan (city) level and turn off when zoomed out. Features can also be turned off when zoomed in, below a scale.

Set visibility ranges for labels

The project has spatial bookmarks you'll use to get appropriate map scales for your specific screen.

1. Open **Tutorial2-8.aprx** from **Chapter2\Tutorials** and save the project as **Tutorial2-8YourName.aprx**.

   The project opens with all layers on but no labeling. You'll set a visibility scale to automatically turn off labels when zoomed below a selected scale.

2. Zoom to the **West Village** bookmark.

   At this map scale and zoomed in, you'll set the visibility range for zoning and land-use labels so they don't display when zoomed out beyond the current scale.

3. In the **Contents** pane, click **ZoningLandUse**.

4. On the **Labeling** tab, in the **Layer** group, click **Label**.

5. In the **Visibility Range** group, for **Minimum Scale**, click **<Current>**.

6. Zoom out and in to see that the **ZoningLandUse** labels are on only when zoomed to the **West Village** bookmark or closer.

> **YOUR TURN**
>
> Similarly, set the visibility range of the **Schools** and **Water** labels to turn off when zoomed out beyond the scale of the **West Village** bookmark. Zoom to the full extent.

Set visibility ranges for feature layers

You can turn a feature layer on and off using visibility ranges, too. For example, at smaller scales (zoomed out), the school points should be turned off to reduce clutter. Borough features are useful at a full map extent but not needed when zoomed in. When you turn off the features using visibility ranges, labels will also turn off.

1. Zoom to the **Lower Manhattan** bookmark.

   The features at this scale are smaller than they were at the scale set from the **West Village** bookmark. The school and borough features should be visible to users based on the current map scale.

   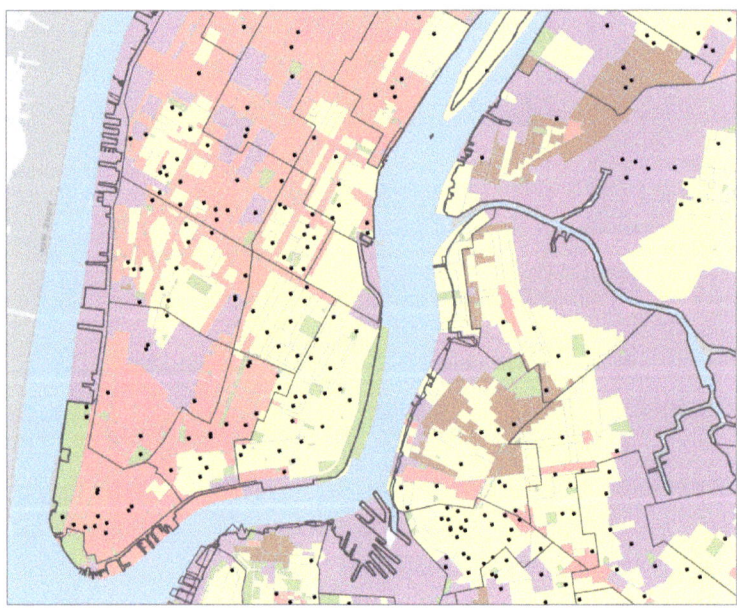

2. In the **Contents** pane, click **Schools**.

3. On the **Feature Layer** tab, in the **Visibility Range** group, click **Minimum Scale** > **<Current>**.

4. Zoom out and in to see the school points disappear and reappear.

5. Zoom to the **Lower Manhattan** bookmark.

6. In the **Contents** pane, click **Boroughs**.

7. On the **Labeling** tab, turn on **Label**.

8. On the **Feature Layer tab**, in the **Visibility Range** group, click **Maximum Scale** > **Current**.

9. Zoom in to see the **Boroughs** boundaries and labels disappear.

> **YOUR TURN**
>
> Zoom to the **Lower Manhattan** bookmark. Set the visibility range so the **Neighborhoods** and **Water** features turn off when zoomed out beyond the **Lower Manhattan** bookmark. Zoom to the full extent. The only layers that should be on are **Boroughs** (with labels) and **ZoningLandUse** (without labels). Save your project.

Assignments

This chapter has assignments to complete that you can download with data from ArcGIS Online at links.esri.com/GISTforPro3.4Assignments.

CHAPTER 3

Maps for end users

LEARNING GOALS

- Build map layouts and charts.
- Share maps in ArcGIS Online.
- Use Map Viewer in ArcGIS Online.
- Use ArcGIS StoryMaps℠.
- Use ArcGIS Dashboards.

Introduction

In chapter 2, you learned how to make maps in ArcGIS Pro. This chapter is about sharing those maps with people who don't have ArcGIS Pro or GIS skills beyond map navigation. For example, you may work on a project team that needs to include maps as part of a report, but the intended customer has only limited GIS skills. After finishing this chapter, you'll be able to provide maps in any of the following formats:

- **Reports, slides, posters, or static maps for websites:** Create static maps with map surrounds, including legends, titles, scale bars, and north arrows (if the top of the map isn't north). Create corresponding charts from map data.
- **Online interactive maps:** Share your maps in ArcGIS Online and include them in apps such as ArcGIS Field Maps, ArcGIS StoryMaps, or ArcGIS Dashboards.
- **Interactive online reports and presentations:** Use your ArcGIS Online maps in an ArcGIS StoryMaps story that includes text, images, videos, credits, and other content in a structured format. A story is self-contained and reads like a book, top to bottom. Also use the same maps and other content in an ArcGIS StoryMaps briefing that has slides for making presentations. Share the URL of your story or briefing with others.
- **Interactive online dashboard:** Use your ArcGIS Online maps in Dashboards and monitor spatial data over time with interactive maps, tables, charts, and other elements.

Tutorial 3-1: Building layouts and charts

In this tutorial, you'll build a map layout that has two maps. Once you finish the tutorial, you'll be able to build any kind of layout, with one, two, or more maps. The purpose of the layout you'll build next is to provide state-level information on the location of jobs in the arts, plus corresponding average annual wages.

The layout you'll build doesn't have a layout title (such as "Maps Showing Arts Employment"). The reason is that this layout, like most layouts you'll build, is destined to be used as a figure in a report, a slide in a presentation, or an image on a website. The layout title is better created in a word processer as a figure caption, in a presentation package as a slide title, or on a website to enable matching of the fonts and styles of other figures used in those formats.

The following image is the finished layout in edit mode, the mode in which you create and modify a layout. Notice the blue guidelines that you'll create. Maps and other elements of the layout snap to guidelines, allowing you to place the elements easily and precisely.

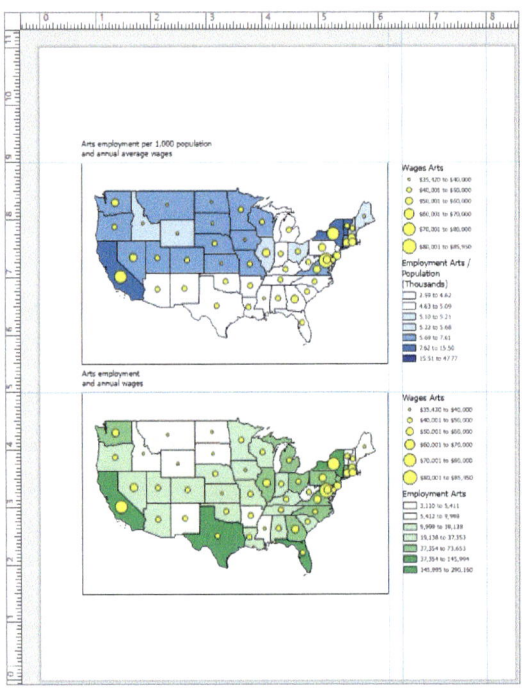

Set up the Tutorial 3-1 project

A layout starts as a blank canvas to which you can add one or more maps, map surrounds (such as legends and map scales), and other elements of your choice and design. Although it's possible to include tables and charts within a map layout, if you're developing a report, presentation, or website, it's better to keep tables and charts separate from the map and make them stand-alone figures. This helps keep your layouts simple and clear. Both maps used in this layout already exist and are ready for use in the Tutorial 3-1 project file you're about to open.

Create a layout and add maps to it

1. Open **Tutorial3-1.aprx** from **Chapter3\Tutorials** and save the project as **Tutorial3-1YourName.aprx**.

   The two available maps are **Arts Employment per 1,000 Population** and **Arts Employment**.

2. On the **Insert** tab, in the **Project group**, click **New Layout**, and under **ANSI – Portrait,** click **Letter 8.5″ × 11″**.

   You may have to click the **Project** button or make your ArcGIS Pro window wider to see the **New Layout** button.

3. In the **Catalog** pane, expand the **Layouts** heading and rename the layout **Arts Employment Layout**.

4. On the **Insert** tab, in the **Map Frames** group, click **Map Frame**. Locate the **Arts Employment per 1,000 Population** map and click **Default Extent**.

5. Click and drag a bounding box in the top half of the layout, leaving some border, and release the mouse.

   This adds the selected map to the layout.

6. Likewise, insert the **Arts Employment** map in the bottom half of the layout.

   You'll resize and relocate these maps in the following section.

Resize and place the two maps

Next, you must resize the two maps to be the same size with matching borders. Planning and trial and error are involved in getting a layout to look right. This section, however, provides dimensions and placement locations to save you time and ensure your results match those of the finished layout shown at the beginning of this tutorial.

1. In the layout, right-click the **Arts Employment per 1,000 Population** map (top map with blue color scheme) and click **Properties**.

2. In the **Element** pane, click the **Placement** button and type **5.5** for **Width** and **3.5** for **Height**.

3. Modify the second map (bottom map, green color scheme) using the same dimensions.

4. Select one of the maps. On the **Layout** tab, in the **Map** group, click **Full Extent**. Click the **Fixed Zoom In** button once so that the map fills the frame.

5. Repeat step 4 for the other map.

Add guides and snap maps to the guides

In this section, you'll use guides to position objects in the layout (maps and legends for now).

1. If your layout doesn't have horizontal and vertical rulers, right-click inside the white area of the layout and check the box for **Rulers**. Make sure that **Guides** is also checked.

2. Right-click the vertical ruler at the **5.0**-inch mark and click **Add Guide**. Add a second guide on the vertical ruler above the one you just created. Hover over the new guideline in the white area of the ruler and drag the guide to the **9.0**-inch mark.

   As you drag the arrow, a pop-up indicates the current position so you can be precise in placement.

3. On the horizontal ruler, add guides at **6.25**, **6.5**, and **8.0** inches.

   You'll snap the legends and maps to these guides.

4. Without resizing it, drag the top map and snap its upper-right corner to the intersection of the **9.0** horizontal and **6.25** vertical guidelines.

   Check the figure at the beginning of the chapter to see this placement.

5. Without resizing it, drag the bottom map and snap its upper-right corner to the intersection of the **5.0** horizontal and **6.25** vertical guidelines.

Insert legends

You'll build a legend for all layers that have been turned on (all layers are already turned on, in this case) for each map. The legends are dynamic: if you change the map symbology, turn layers on or off, or make other changes, the legend in the layout automatically updates. You will insert the improved legend into your layout and simplify the map legend.

A major principle of graphic design, according to mathematician John Tukey, is to "minimize ink!" If a graphic element does not add value, don't use it or delete it. You'll put that principle to work in the following steps for the legends.

1. Click the top map to select it.

2. On the **Insert** tab, in the **Map Surrounds** group, click the **Legend** arrow and click **Legend 1**. Drag a rectangle that snaps to the rectangle that your guides have formed to the right of the map.

   Check the figure at the beginning of this chapter to see this placement. ArcGIS Pro creates and draws the legend.

3. Click the bottom map to select it and similarly create its legend.

4. Right-click the top legend and click **Properties**.

5. In the **Element** pane, under the **Legend** tab, click the **Options** button.

6. Under **Legend**, uncheck the **Show** box for **Title**.

7. For **Legend Items**, click **Show properties**.

8. Under **Show**, uncheck the box for **Layer name**.

   *Tip: If you make a mistake so that your legend disappears, click the **Undo** button on the top menu.*

9. Select the bottom legend and repeat steps 5 through 8.

Insert text

Next, you'll insert text, titling each map (but not the whole layout).

1. On the **Insert** tab, in the **Graphics and Text group**, select the **Rectangle Text** option from the gallery.

2. Starting above the upper-left corner of the top map, create a title-sized rectangle.

3. In the **Rectangle Text** box, type **Arts employment per 1,000 population**, press **Enter**, and type **and annual average wages**.

> **YOUR TURN**
>
> Insert text for the bottom map, using the preceding steps. Type **Arts employment**, press **Enter**, and type **and annual average wages**. Save your project.

4. In the **Catalog** pane, expand **Layouts**, right-click **Arts Employment Layout**, and click **Export to File**.

5. In the **Export Layout** pane, change the **File Type** to **JPEG** and click **Export**. Your exported layout should show two maps, each with a title and a legend.

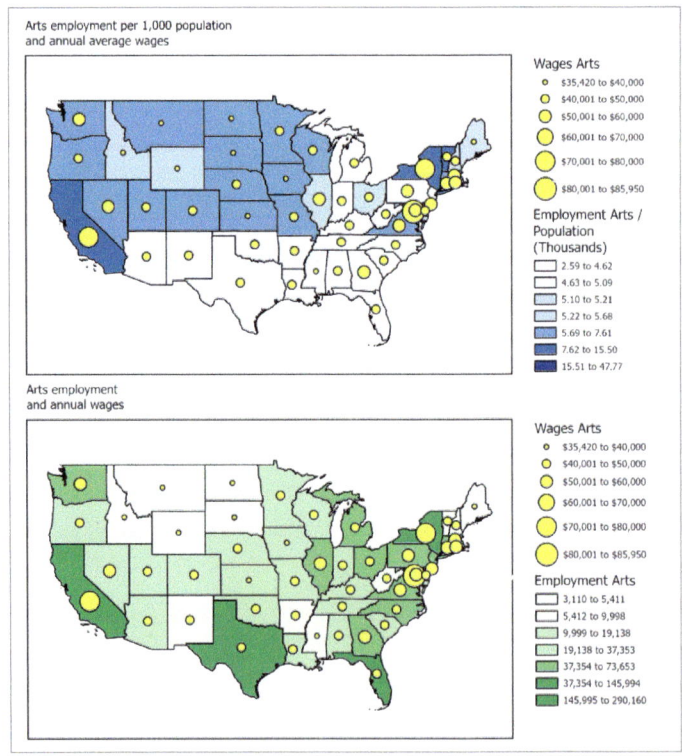

Create charts

1. Click the **Arts Employment** map tab.

2. In the **Contents** pane, right-click the **Employment** layer and click **Create Chart > Bar Chart**.

   The **Chart Properties** pane opens on the left.

3. In the **Chart Properties** pane, make the following selections:
   - **Category or Date**: Name
   - **Aggregation**: <none>
   - **Numeric field(s)**: Select > Employment Arts
   - **Sort**: Y-Axis Descending

4. In the chart, drag your mouse over the first 10 bars (**California** to **Massachusetts**) to highlight them.

5. At the top of the chart, for **Filter**, click the **Selection** button.

   These actions select those bars and then limit the chart to only them.

6. In the **Chart Properties** pane, click the **General** tab and apply the following settings:
   - **Chart Title**: Arts Employment for the Top 10 States
   - **X-Axis Title**: State
   - **Y-Axis Title**: Employment

   Tip: If the chart is intended for a report, you have the option to leave the **Chart title** blank by unchecking the box.

7. Click the **Series** tab. Click the **Symbol** for **Employment Arts** and change it to a color of your choice.

8. Click the **Axes** tab. Change the **Label character limit** to **13** (allows Massachusetts to be fully displayed as a bar label) and set the **Interval** to **50000**.

9. Click the **Format** tab. Click the **Text Elements** button and select **Chart Title**. At the bottom of the pane, set the **Font Size** to **12 pt**. Select **X-axis Labels** and **Y-axis Labels** and set the **Font Size** to **10 pt**.

10. Close the **Chart Properties** pane. Adjust the height of the **Chart** window so that the bars on the chart are at the height you prefer. If you prefer to make it smaller, undock the window and adjust its size.

11. At the top of the chart, click **Export** > **Export As Graphic**. Save the chart to this chapter's resources folder. Set the **Name** to **StateArtsEmployment** and the **File Type** to **JPEG**. Click **Save**.

    Your chart is included in the **Contents** of your map under **Charts**. If needed, you can activate your chart and edit it from **Contents**.

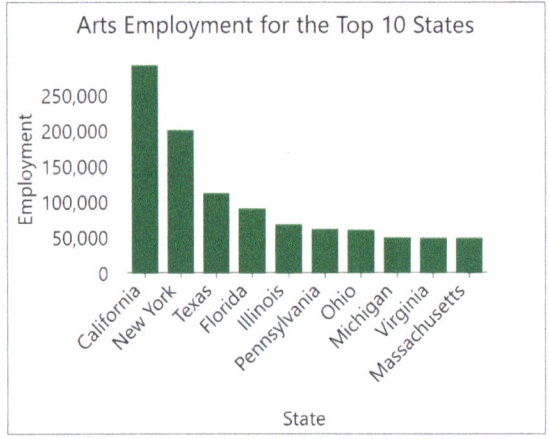

> **YOUR TURN**
>
> On the **Arts Employment** map, clear the selection (top 10 states for employment). For the same layer, **Arts Employment**, create a **Scatter Plot**. The x-axis should be **Population (Thousands)** and the y-axis should be **Employment Arts**. When finished, you'll see that two points are far above the resulting trend line, meaning they have exceptionally high employment in the arts relative to the norm. Save your project.
>
>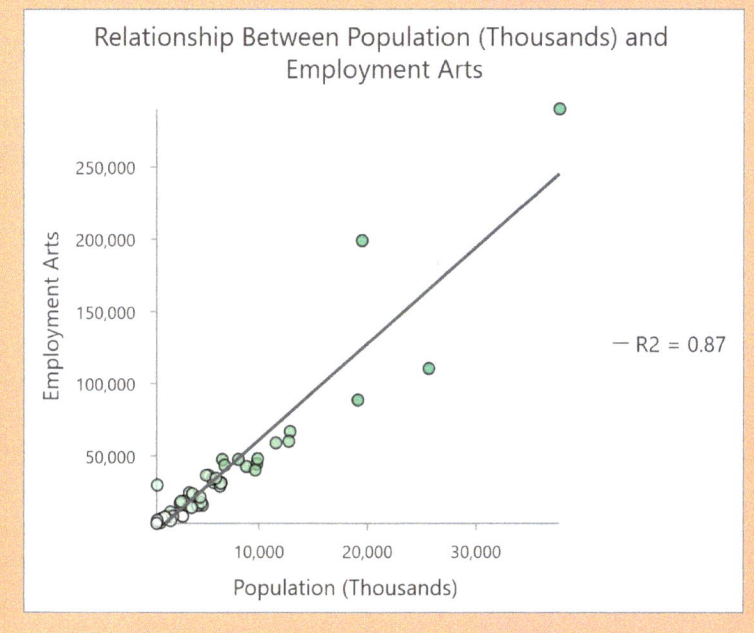

Tutorial 3-2: Sharing maps online

In this tutorial, you'll share ArcGIS Pro maps, publishing them in ArcGIS Online as web maps. Your username and password for ArcGIS Pro also work for signing into ArcGIS Online in your browser. After your maps from tutorial 3-1 are online, you'll do additional work on them using ArcGIS Online Map Viewer, which has functionality similar to ArcGIS Pro for working with and symbolizing maps. Later in this chapter, you'll use the web maps as the basis for a story in ArcGIS StoryMaps and for slides in ArcGIS StoryMaps Briefings.

Set properties for online sharing

To share or publish a map online, you must check a property of the map. You must also have a basemap layer, which is already the case for both maps, the **Light Gray Canvas** basemap.

1. Open **Tutorial3-2.aprx** from **Chapter3\Tutorials** and save the project as **Tutorial3-2YourName.aprx**.

   This tutorial uses two of the maps from tutorial 3-1 but with additional feature classes for metropolitan areas, including a cost-of-living index, and visibility ranges so that the metropolitan areas display only when zoomed in close enough.

2. In the **Arts Employment** map, zoom in to see the **Metropolitan Employment** feature layer and then zoom to the full extent.

3. At the top of the **Contents** pane, right-click **Arts Employment** and click **Properties**. Under **General**, confirm that the box for **Allow assignment of unique IDs for sharing web layers** is checked. Ignore the resulting warning and click **OK**.

   Each layer you share is automatically assigned an ID, based on its drawing order in the **Contents** pane. This is necessary for ArcGIS Online.

> **YOUR TURN**
>
> Repeat steps 2 and 3 of the preceding section for the **Cost of Living Index** map.

Share maps in ArcGIS Online

You will publish the maps you create in ArcGIS Pro to ArcGIS Online. You have two maps to share: **Arts Employment** and **Cost of Living Index**. Once online, you can use your shared maps in a variety of web-based apps.

1. Click the **Arts Employment** map. On the **Share** tab, in the **Share As** group, click **Web Map**.

2. In the **Share As Web Map** pane, under **Item Details**, set the **Name** to <mark>Arts Employment \<your name\></mark>.

3. In **File Explorer**, browse to **Chapter3\Tutorials\Resources** and open **MapSharing.pdf**.

   This document has text for you to copy and paste into ArcGIS Pro when you share your two maps.

4. To fill in the **Summary** text box, copy and paste the **Summary 1-1** text from **MapSharing.pdf**.

5. To fill in the **Tags** text box, copy and paste **Tags 1-2**.

6. Under **Sharing Level**, click **Everyone (public)**.

   You can also share only with your organization or with a specified group from your organization instead of with everyone.

7. Click **Analyze**. Ignore the warning about display filters.

8. Click **Share**.

   Your map is shared and available online.

> **YOUR TURN**
>
> Similarly, share the **Cost of Living Index** map, adding your full name and copying and pasting text from **Summary 2-1** and **Tags 2-2** from **MapSharing.pdf**. Save and close ArcGIS Pro.

Use Map Viewer in ArcGIS Online

You'll open your maps in ArcGIS Online Map Viewer, change symbology, and configure pop-ups. In an optional step, you can download ArcGIS Field Maps on your mobile device and view your maps in it.

1. Go to <mark>arcgis.com</mark> and sign in with the same username and password you use for ArcGIS Pro.

2. Click **Content** > **My content**.

   You can see your two web maps and accompanying files.

3. Click the **Arts Employment <Your Name>** web map.

   The item's detail page with a summary and properties of the web map opens.

4. Click **Open in Map Viewer**.

   The viewer opens with your map and menus for working with the map.

5. In the bottom-left corner, click the **Expand** button in the lower left to open the **Contents** (dark) toolbar. In the bottom-right corner, click the **Expand** button to open the **Settings** (light) toolbar.

   You'll use the **Contents** toolbar to manage and select map components and the **Settings** toolbar to configure them.

6. If **Layers** is not highlighted on the **Contents** toolbar, click it.

   The **Layers** pane lists all layers in the map. You can select a layer in the **Layers** pane and then use the **Settings** toolbar to modify the selected layer.

Change the style of a layer

You can symbolize layers in Map Viewer just as you can in ArcGIS Pro. In this section, you'll change the color of a layer to get a quick look at symbology in Map Viewer. You can explore the full range of symbology options on your own.

1. In the **Layers** pane, click **State Wages**.

2. In the **Settings** toolbar, click **Styles**.

   The left of the **Settings** toolbar becomes the **Styles** pane.

3. In the **Styles** pane, under **Pick a style**, click **Counts and Amounts (size)**.

   This symbology was applied in ArcGIS Pro. Map Viewer recognizes the symbology and applies the corresponding style.

4. In the **Style options**, click the symbol under **Symbol style**.

The **Symbol style** pane opens on the left.

5. Click **Color** and choose an orange color. Click **Done**.

6. In the **Style options** pane, click **Done** and **Done** again.

   The **Styles** pane closes and the map updates with the color you selected.

Configure pop-ups

1. Zoom in and pan to Southern California, including Los Angeles, so that the **Metropolitan Employment** point layer becomes visible.

2. In the **Layers** pane, click **Metropolitan Employment**.

3. On the **Settings** toolbar, click **Pop-ups**.

4. In the **Pop-ups** pane, turn on **Enable Pop-ups**.

5. Click **Fields list** and click **Select fields**. Click **Select all** and then click **Deselect all**.

   All fields in the layer are deselected, and you can choose the fields you want to display.

6. Click the following fields to select them for display in pop-ups for the **Metropolitan Employment** layer:
   - **Employment Arts**
   - **Wages Arts**

7. Click **Done** and close the **Pop-ups** pane.

8. On the map, click the point for **Los Angeles** to see its pop-up. After viewing the pop-up, close it.

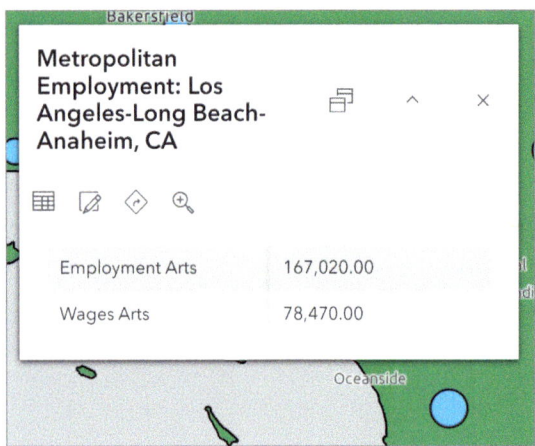

9. In the lower right of the map, click the **Default map view** button, and click the **Zoom In** or **Zoom Out** buttons as needed so that the map fills the available space.

10. On the **Contents** toolbar, click **Save and open** > **Save** to save your map.

> *Important note:* The current extent of the map becomes the default for display whenever you save the map. So be careful to zoom to full extent the last time you save a map during a session.

YOUR TURN

On the **Arts Employment** map, create pop-ups for the **State Employment** layer. Include the following fields in the pop-up: **Population**, **Employment Arts**, and **Wages Arts**. Click the **Default map view** button, click the **Zoom In** or **Zoom Out** buttons as needed so that the map fills the available space, and save your map.

11. On the **Contents** toolbar, click **Save and open** > **Open map**. Open the **Cost of Living Index** map.

> **YOUR TURN**
> Apply the skills you've learned to this map, creating a pop-up for the **State** and **Metropolitan Cost of Living Index** layer using the Index field for both state and metropolitan areas. Set the extent and save your map.

> **YOUR TURN** (Optional)
> Browse to either the App Store or Google Play and search for and install the free app ArcGIS Field Maps. You can sign in with the same username and password you use for ArcGIS Pro and ArcGIS Online, or you can proceed without signing in and search for maps. If you sign in, all your ArcGIS Online maps are quickly available, including any you haven't made public. All the interactive navigation and other functionality you'd expect for map use are easy to find.

Tutorial 3-3: Creating a story and a briefing in ArcGIS StoryMaps

ArcGIS StoryMaps allows you to create stories that include text, web-based interactive maps, images, videos, and other content. A story is intended to be read by individuals, with paragraphs and corresponding displays in a logical flow. ArcGIS StoryMaps also allows you to create briefings that consist of a series of slides with bulleted talking points, interactive maps, and other content for a presentation to an audience. Once published online, you can share the URL of your story or briefing with anyone.

The story and briefing that you'll create in this tutorial use the **Arts Employment** and **Cost of Living Index** maps that you shared online from tutorial 3-2.

Create a story

1. In a browser, go to **arcgis.com** and sign in.

2. Click **Content** and click **Create app**. Select **ArcGIS StoryMaps**.

Throughout this tutorial, you'll include content from **Chapter3\Tutorials\Resources**. You'll copy and paste text and insert charts and images in your story. Each section for copying and pasting from **StoryMapManuscript.pdf** is numbered for reference.

Tip: It's good practice to develop and write content in a word processor, and then copy and paste it when you're ready to author your story or briefing. This allows you to concentrate on technical issues and design when working on the web and be confident that the content is stable. Also, if you're working on a team with more than one member writing text, you can share and revise each other's work in a document first.

3. From **Chapter3\Tutorials\Resources\StoryMapManuscript.pdf**, copy and paste **Title 1-1** and **Subtitle 1-2** into their respective locations on the blank cover page. Change **Your Name** to **your full name**.

   As you add new content or make changes, ArcGIS StoryMaps saves your work automatically. Currently, your story is classified as a draft because you haven't published it yet to the web.

   The title becomes the file name for your story in ArcGIS Online's **Contents**. It's important that you use your full name in the title to make the resulting file name unique if others use the same overall ArcGIS Online account, such as in a class or organization.

4. Scroll up and click **Add cover image or video**. From **Chapter3\Tutorials\Resources**, add the **pexels-engin-akyurt-6137963.jpg** image.

   You can find alternative images from several sites, including pexels.com, which doesn't require giving credit for authorship. Nevertheless, save the URL where you get your image so that you can still give credit in the **Credits** section of your story later. The image used in this example was downloaded from www.pexels.com/photo/light-red-art-blue-6137963.

5. In the story builder header, click **Design**. Under **Cover**, click **Full** (third cover design). Close the **Design** panel.

6. Click the **Change panel appearance** button (palette and brush icon). Under **Horizontal** position, click the **Center aligned** button to align the text in the center.

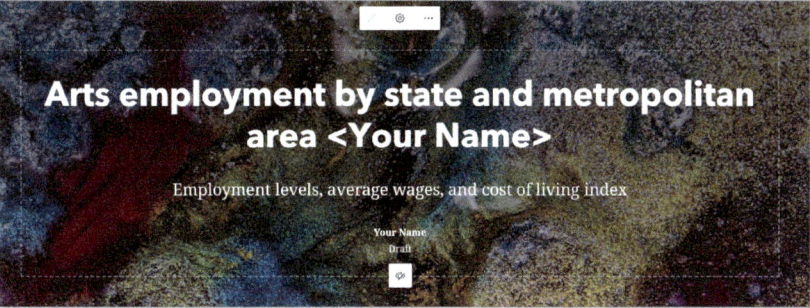

7. Click the **Settings** button (gear icon). In the **Image options**, click the **Properties** tab and enter the attribution in the **Attribution** field. If you used the image provided in the Resources folder, enter the following link: **www.pexels.com/photo/light-red-art-blue-6137963**.

8. For the **Alternative text** field, copy and paste **Alternative Text 1-3** from **StoryMapManuscript.pdf**, and click **Save**.

   This text is available for people with impaired vision. Search engines can also access this text.

   Notice on the story builder header that there are **Undo** and **Redo** buttons to use if you make any mistakes.

Add blocks for the introduction

Each separate element of content in a story is called a "block". Examples of blocks are headings, paragraphs, bulleted lists, images, and so on. Each time you need to add content of a different type, you'll open the block palette, choose the needed block type, and add the content. Generally, you'll add blocks in the order they would appear in your story, but you can always insert new blocks where needed or move blocks up or down.

1. Scroll down and click the **Add content block** button.

   The block palette opens.

2. Click **Text**. From **StoryMapManuscript.pdf** and copy and paste **Heading 2-1** to your story.

   The next step is key to your understanding of configuring blocks. You'll configure all blocks in a similar way.

3. Highlight the text to select it. Click **Paragraph** and change the style to **Heading 1**.

   Later in this tutorial, you'll add navigational links at the beginning of the story so readers can quickly navigate to headings that interest them. You should design headings to be meaningful and consistent.

4. Below **Heading 1**, click the **Add content block** button and click **Text**. From **StoryMapManuscript.pdf**, copy **Subheading 2-2**. Paste the text and change the style to **Heading 2**.

5. Below **Heading 2**, add another **Text** block. From **StoryMapManuscript.pdf**, copy **Paragraph 2-3**. Paste the text and keep it styled as **Paragraph**. Add breaks between the paragraphs.

6. Select the last sentence of the second paragraph starting with "**You might be better off . . .**" and click the **Italic** button.

   The sentence is now in italic.

   ## Arts Employment

   ### Where are the jobs in the arts field? What are average wages versus cost of living?

   If your career interests lie in the arts field, broadly speaking (including arts, design, entertainment, sports, and media occupations), this story map can help you analyze locations across the U.S. where you might live in regard finding jobs and financial prospects.

   Of importance for job searching in the arts field is that there are huge variations by state and metropolitan area in employment levels, average wages, and cost of living. As you might suspect, wages in the arts and cost of living are positively correlated (71%), so the job with the highest wages across different cities of the country may not be the best option financially because of high cost of living. *You might be better off with a good, but not top-paying job elsewhere because of lower cost of living.*

   To help you explore metropolitan areas for arts employment, this story map has two interactive maps that provide data on employment levels, average annual wages, and a cost of living index by state and metropolitan area. Also included is a sample calculation showing how to use the data from the maps to analyze the financial prospects for you of different metropolitan areas

7. Copy **Subheading 2-4**, paste it in another **Text** block, and style it as **Heading 2**.

8. Copy **Paragraph 2-5** and paste it in another **Text** block.

Add a sidecar block with a map

A default sidecar block with a docked panel has media (in this case, your interactive employment map) in a panel on the right and text and other content (for example, charts or tables) in a scrolling panel on the left. The docked panel is useful when you have a lot of narrative to enter in the left panel, as is the case with the employment map. The floating panel option is for cases with short narrative (a caption or a few sentences) and maximizes the size of the media.

1. Click the **Add content block** button, scroll down, and click **Sidecar**. Click **Docked** and click **Save**.

2. Click **Add**, and from the list, click **Map**. Select your **Arts Employment** map. In the left panel, click the **Options** button and enable **Allow map navigation**, **Search**, **Legend**, and **Scale bar**. Click **Save**.

3. Try out the available map functionality, including the legend (circle in lower left of the map). Zoom in so that the **Metropolitan Employment** layer appears. Try the pop-ups. Finish by making your map fill the available window.

4. Click the map's **Options** button, paste **Alternative Text 2-6**, and click **Save**.

Add content for the left panel of the map

1. In the **Sidecar** block, on the left panel, paste **Heading 3-1** as a **Text** block and style it as **Heading 1**.

2. Paste **Subheading 3-2** in another **Text** block and style it as **Heading 2**.

3. Paste **Bulleted List 3-3** in another **Text** block and style it as **Bulleted List**. Place the pointer at the start of each sentence that you think should be bulleted and press **Enter** to create bulleted items.

4. Below the list, click the **Add content block** button and click **Image**.

5. From **Chapter3\Tutorials\Resources**, upload **Table1.png**. Click **Add**.

6. In the **Add a caption** prompt below the inserted image, paste **Caption 3-4**.

> **YOUR TURN**
>
> Add **Subheading 3-4** as **Heading 2**, **Table2.png** as **Image**, and **Caption 3-5** as the table's caption.

Enable navigation

Next, you'll add heading links to the parts of your story.

1. In the story builder heading, click **Design**.

2. Turn on **Navigation** and **Credits** and then close the panel.

   Below the cover image, headings from your story now appear as links in a horizontal navigation bar.

3. Try the links.

Add credits

1. Scroll to the bottom of your story, as far as you can go.

   The **Credits** section appears. As is the case for anything you author, you must add credits to anything that uses the work of others.

2. For the heading, type **Credits**. Skip the credits description.

   This heading will also now appear as a link in the navigation bar.

   In **StoryMapManuscript.pdf**, under the **Credits** section, there are four credit entries. Each entry has two parts: text for the **Content** (first part of the entry) and text for the **Attribution** (second part of the entry).

3. Paste the text for the **Content** and then the **Attribution**, including the URL. Click the **Add credit** (plus) button as needed to add additional credit lines.

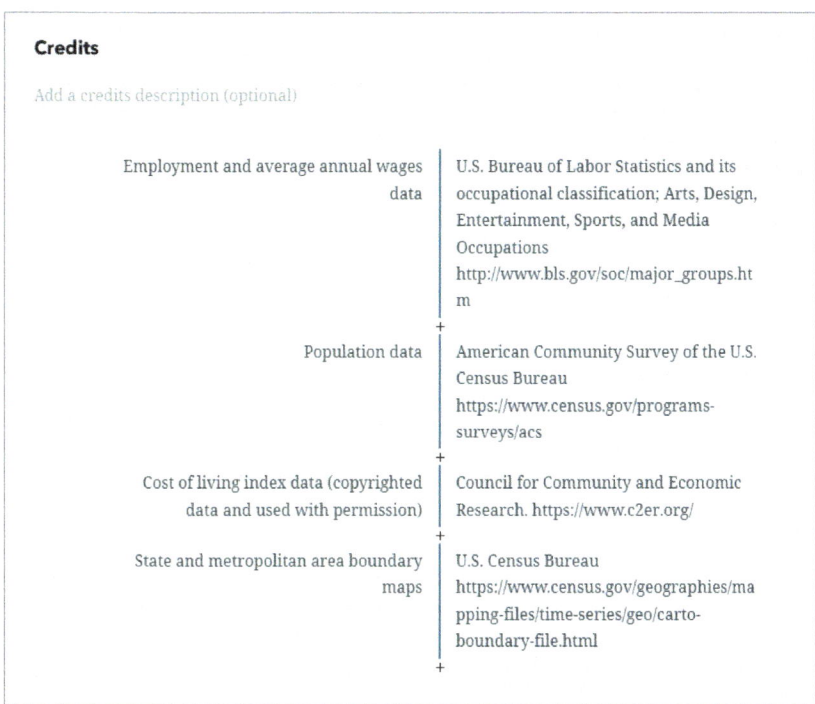

Preview and publish your map

Next, you'll preview your story as it would appear on a smartphone, tablet, or computer monitor and publish it to the internet.

1. In the story builder heading, click **Preview**.

2. On the bottom right, try the different preview options on desktop, tablet, or phone. When you have finished, close the preview.

3. Click **Publish**. Set the **Share** level to **Everyone (Public)** and click **Publish**.

4. If you get a **Content authentication needed** warning, click **Yes, publish anyway**.

5. After the story publishes, it opens in your browser.

   Your story is listed under **Content** in ArcGIS Online and can be edited from there. Click the **Options** button at the end of the story's listing and then click

Edit Story. Also, click the name of the story in **Content**, scroll down, copy the URL for the story, and open it in a browser.

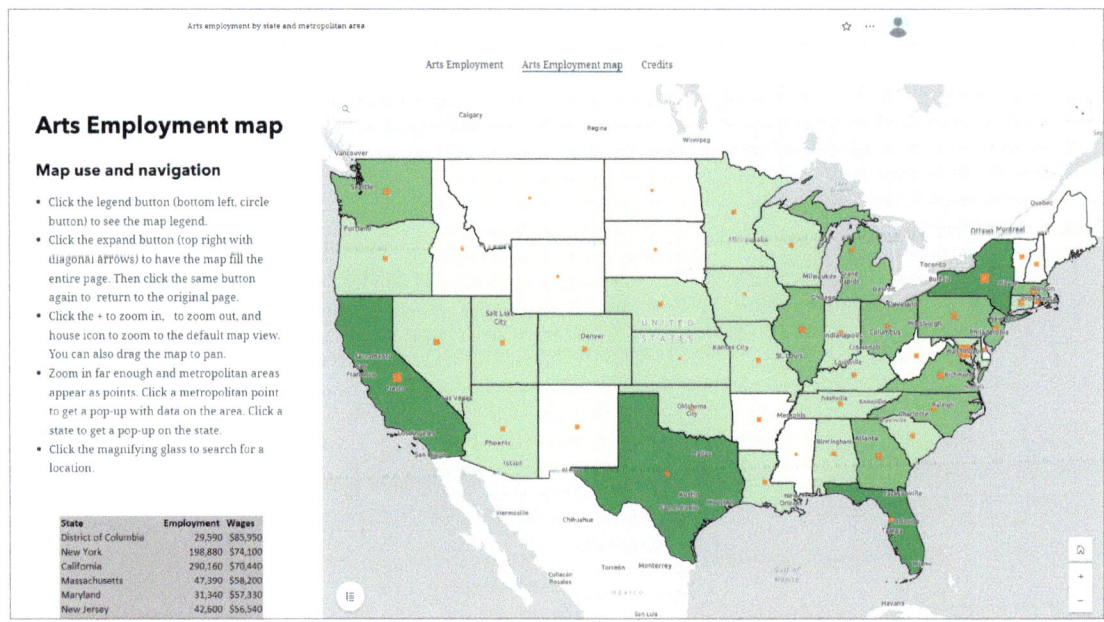

YOUR TURN

Add a second map to your story. At the bottom of your story in the **Sidecar** panel, click the **New slide** button (circle with a plus sign). Add the **Cost of Living Index** map to the right panel and turn on the same **Options** settings as the previous map. Add content to the left panel using text from section 4 of **StoryMapManuscript.pdf**. Publish your story when finished.

Create a briefing

1. In ArcGIS Online, go to **Content**, click the **App Launcher** (nine dots), and select ArcGIS StoryMaps. Click **Briefings** > **Create new briefing**.

2. From **Chapter3\Tutorials\Resources\BriefingsManuscript.pdf**, copy and paste **Title 1-1** and **Subtitle 1-2** into their respective locations in the blank **Title** slide.

3. From the same folder, add **pexels-engin-akyurt-6137963.jpg** for the cover image in the **Title** slide.

4. Click the **Add slide** button and select **Single Panel with Title**.

5. Copy and paste slide 2 content. Format the **Body** as a **Bulleted List** and make three bullet points, one for each type of art field.

6. Add another **Single Panel with Title** slide. Copy and paste slide 3 content. For the **Body**, create bullets under each bolded introductory phrase.

7. Add a **Media only** slide and add your **Arts Employment** map.

8. Add another **Media only** slide and add your **Cost of Living Index** map.

9. Preview your briefing and then publish it.

    As usual, you can click your briefing in **Content** and copy its URL to open it in a browser and share with others.

Tutorial 3-4: Creating a dashboard in ArcGIS Dashboards

Dashboards generally are tools used by organizations to allocate resources in response to changing demands for goods or services over time. Dashboards are visual displays of spatial data feeds over time in an easy-to-understand format, with a map that has updated data as the center of focus along with tables and charts. You'll create a dashboard using ArcGIS Dashboards.

Most cities have phone and web-based systems for receiving nonemergency (311) service requests from residents, such as needs for sidewalk or road repairs. In this tutorial, you'll create a dashboard for displaying Pittsburgh's 311 service requests to remove debris and overgrowth from properties. The policy for prioritizing requests is to respond to the oldest requests first (first in, first out). At the same time, to minimize time wasted traveling between jobs, the city works manager assigns ground crews to fulfill all removal requests in the immediate area of priority requests. This dashboard allows the manager to zoom to an area that needs service and get information to assign work crews.

The image shows the finished dashboard you'll create with map, table, and charts. The map is zoomed to a portion of the **Spring Hill – City View** neighborhood of Pittsburgh. In the map, the older the request for service (the **Age** column, measured in days), the larger the corresponding map symbol. The dashboard is dynamic, configured to automatically show only service request data in the table of the observed, zoomed-in map extent. Similarly, the charts show only data from the observed map extent. The requests for service on the map all have **Status** of 0, meaning that service hasn't yet been provided. When assigned for service or when service is completed, the manager changes **Status** to **1** or **2**.

The map you will use for this tutorial is already prepared, ready for you to share in ArcGIS Online. The dashboard itself is easy to build using the Dashboards app. When adding a new element to the dashboard, you must first select a position for it on the border of the dashboard window. Then you choose the element type (map, table, and so forth) and configure it.

Note that, in practice, the IT department of an organization, such as Public Works, would have to automate the data feed to a dashboard (a task beyond the scope of this book). Nevertheless, a person or team with your GIS and problem-solving skills would design and build the dashboard just as in this tutorial.

Open the Tutorial 3-4 project

1. Open **Tutorial3-4.aprx** and save the project as **Tutorial3-4YourName.aprx**.

    The map is ready for publishing (sharing) to ArcGIS Online. Included are all outstanding, unmet service requests. As you can see, requests in the northwestern part of Pittsburgh have been ignored and should get priority.

2. Try the pop-ups for the **DebrisOvergrowth** layer. Also, zoom in far enough to see the labels for the **Neighborhoods** and **DebrisOvergrowth** layers.

The labels for the **DebrisOvergrowth** layer show the IDs for each record. The dashboard user needs these IDs when assigning ground crews. The 311 data source didn't include the street address for requests (which were collected but not provided) but instead provided (x,y) coordinates. In an actual dashboard for the ground crew manager, the street address would be included to locate worksites for ground crews.

3. Zoom back to full extent.

> **YOUR TURN**
>
> Share the **311 Debris Overgrowth** map to your ArcGIS Online account as a **Web Map** (refer to the beginning of tutorial 3-2 for the needed steps). In the map and its layers, all needed properties are set for sharing. Be sure to add your full name to the web map name in the **Share as Web Map** pane. Share the map with everyone. Save the map in ArcGIS Online at full extent.

Create a dashboard and add a map element

At any point in creating a dashboard, you can save your work and close. To start working on a dashboard again, return to **Content** in ArcGIS Online and click the dashboard's name to edit.

1. In a browser, sign in to your ArcGIS Online account.

2. In the upper right of your screen, click the app launcher and click **Dashboards**.

3. Click **Create dashboard** and apply the following settings:
   - Title: **Ground Crew Dashboard <Your Name>** (include your full name to make the title, which also serves as the dashboard file name, unique)
   - Tags: **311 calls**, **Debris overgrowth**, **Crew assignments** (press **Tab** after typing each tag)
   - Summary: **Dashboard for assigning ground crews to remove debris and overgrowth**

4. Click **Create dashboard**.

5. In the bottom-left corner, click the **Expand** button to expand the dashboard toolbar.

6. Click **Add element**. Click **Map** and select your **311 Debris Overgrowth** map.

7. Turn on all available map components, except the **Scalebar** and **Find my location**. Change the **Point zoom scale** to **500**.

8. Click **Done**.

   Your map is now the only element of your dashboard, but you'll soon add other elements. At the top and bottom right, review the map components you turned on.

9. On the dashboard toolbar, click **Save** > **Save** to save your dashboard (unsaved work is denoted by a blue dot on the **Save** button).

10. Try the pop-ups and view the labels in your map.

11. In the upper-right corner of the dashboard, click the **Bookmarks** element. Select the **Spring Hill – City View** bookmark and observe the extent. Click **Initial View**, which is full extent if you shared the map at full extent.

Add a table element

Next, you'll add the map's attribute table as an element in the dashboard. The manager needs the map and table to make assignments to the ground crews.

1. On the dashboard toolbar, click **Add element**. On the right side of the dashboard, click **Add element here** button and click **Table**.

2. In the **Select a layer** window, click **DebrisOvergrowth**.

3. In the **Table** pane, under **Data**, make these selections:
   - **Table Type**: Features
   - **Value fields**: RequestID, RequestDate, RequestType, Age, X, Y
   - **Sort By**: RequestDate, Sort Descending

   This selection allows the manager to see the latest requests at full extent, which helps track the workload. When zoomed to a work area, after you make the dashboard dynamic, the manager can see all the corresponding requests in that area only, old and new.

4. Click the **Table** tab, and for **Size**, click **Small**.

5. Click **Done** and save your dashboard.

Adjust the table

The table is too wide, and some fields are unnecessary. You will adjust the table to delete them.

1. Hover your pointer over the top-left corner of the table and click the **Configure** button (gear icon).

2. In the **Table** pane, delete the **X** and **Y** value fields.

3. Click **Done**.

4. Point between the table and map until the double-headed arrow appears, and then click and narrow the table.

Add a serial (bar) chart

The bar chart you create next will give the manager an idea of the distribution of the ages of requests.

1. On the dashboard toolbar, click **Add Element**. At the bottom of the map, click the **Add element here** button and click **Serial chart**.

2. In the **Select a layer** window, click **DebrisOvergrowth**.

3. In the **Serial chart** pane, under **Data**, make these selections:
   - **Categories From**: Grouped Values
   - **Category Field**: Age
   - **Statistic**: Count

4. Click the **Category axis** tab, and for **Title**, type **Age of request for service (days)**. Change **Title size (px)** to **14**. Expand **Labels** and change **Size (px)** to **14**.

5. Click the **Value axis** tab, and for **Title**, type **Number of requests**. Change **Title size (px)** to **14**. Expand **Labels** and change **Size (px)** to **14**.

6. Click **Done**.

7. Drag the top of the chart down to make it about half as high.

8. Save your dashboard.

> **YOUR TURN**
>
> Add a **Pie chart** element to the right of the **Serial chart** and below the **Table**. From the **DebrisOvergrowth** layer, show the grouped values from the **RequestType** category field using **Count** as the statistic. The pie chart will give the manager information on what type of work is being requested. In the **Chart** tab, set the **Font size (px)** to **14**. Adjust the size of the pie chart as you like. Save your dashboard.

Add interactions to the dashboard

The next steps are critical for the success of the dashboard. When the manager zooms in on the map, all the other dashboard elements should show results for only the zoomed-in map extent.

1. Hover your pointer over the top left of your map and click the **Configure** button.

2. On the **Map actions** tab, click **Filter** and turn on **Pie chart**, **Serial chart**, and **Table**. Click **Done**.

3. Zoom to the **Spring Hill – City View** bookmark.

   The table and charts limit what is displayed based on the change to the map extent.

4. In the upper-right corner of the map, click the **Expand** button.

   If you point to any element on the dashboard, the same button appears.

5. Click the map's **Expand** button.

   The map fills the screen.

6. Click the **Basemaps** button and click through the different basemaps available. Switch back to **Light Gray Canvas**.

7. In the upper-right corner of the map, click the **Collapse** button to return to the full dashboard.

8. Save your dashboard.

Finish your dashboard

1. Above **Add element**, click the three lines and click **Content**.

2. Click your **Ground Crew Dashboard** name and click **Share**. Share your dashboard with your organization or with everyone. Click **Save**.

3. Scroll down and copy the URL for the dashboard.

4. In a new browser tab, paste the dashboard's URL to access it.

5. Zoom to the **Spring Hill – City View** bookmark. Observe the data in the **Table**, **Pie chart**, and **Serial chart**.

6. Close your browser.

Assignments

This chapter has assignments to complete that you can download with data from ArcGIS Online at links.esri.com/GISTforPro3.4Assignments.

PART 2

Working with spatial data

CHAPTER 4

File geodatabases

LEARNING GOALS

- Import data into file geodatabases.
- Modify attribute tables and fields.
- Use Python expressions to calculate fields.
- Join tables.
- Get an introduction to SQL query criteria.
- Carry out attribute queries.
- Aggregate point data to polygon summary data.

Introduction

In this chapter, you'll learn about working with spatial databases and databases in general. A database is a container for the data of an organization, project, or other undertaking for record keeping, decision-making, analysis, and research. A file geodatabase is Esri's simplified database for storing geospatial data, including feature classes and raster datasets, for use by single users or small groups. In terms of data format, a file geodatabase is a file folder that has *.gdb* at the end of its name and contains files. A .gdb is not a file extension but part of the folder name. In ArcGIS Pro, data management and processing in a file geodatabase are done through the **Catalog** pane, tools, and user interface.

For example, **FoodDesertsChicago.gdb** may be the name of a file geodatabase that stores spatial data for analyzing residents' access to grocery stores in Chicago, starting with feature classes for grocery stores, streets, and population in census blocks in Chicago. The workflow for estimating the number of Chicago residents who live in food deserts (say, live a mile or more from the nearest grocery store) would have you create additional feature classes and tables, stored in **FoodDesertsChicago.gdb**.

Although file geodatabases have a simple format, they are powerful spatial data containers. For example, file geodatabases have no practical limits on numbers and sizes of feature classes and raster datasets stored in them. Also, they are optimized for data processing and storage in ArcGIS Pro, and they allow data tables to be related and joined (essential database processes).

You must work tutorials 4-1 and 4-2 sequentially because the work is accumulative. Also note that *attribute*, *field*, *variable*, and *column* are interchangeable terms for the columns of data tables, and *record*, *row*, and *observation* are interchangeable terms for the rows in a data table.

Tutorial 4-1: Importing data into a new ArcGIS Pro project

When you create an ArcGIS Pro project, the software automatically creates a file geodatabase for you as the project's default spatial data container. In this tutorial, you'll create a project and import data from external sources into the new project's file geodatabase.

Create an ArcGIS Pro project

1. Open **ArcGIS Pro**.

2. Under **New Project**, click **Map**. Apply these settings:
   - **Name**: YouthPopulation
   - **Location**: C:\EsriPress\GISTforPro\Chapter4

3. Click **OK**.

   The project opens with a map that uses the **World Topographic Map** and **World Hillshade** basemaps.

4. On the **Project** tab, click **Options**.

   Under **Current Settings**, ArcGIS Pro created the new project, **YouthPopulation.aprx**, as well as a **YouthPopulation.gdb** file geodatabase and **YouthPopulation.atbx** toolbox in the **Chapter4\YouthPopulation** folder. Every new feature class, table, and file you create is saved by default to the project's file geodatabase. You'll learn about creating tools later in the book.

5. Click **OK**.

   Next, you'll save the project file as **Tutorial4-1<YourName>** in keeping with the style of the rest of this book.

6. On the **Project** tab, click **Save Project As**. Under **Project**, click **Folders**, and open the **YouthPopulation** folder. For **Name**, type **Tutorial4-1<YourName>** and click **Save**.

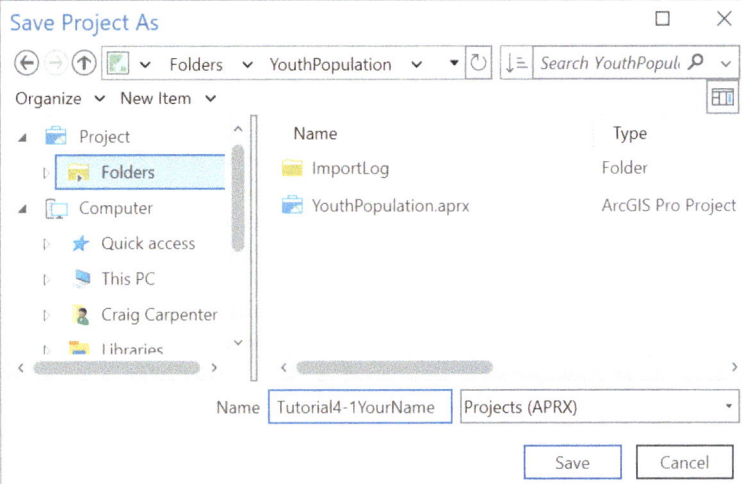

Set up a folder connection

Besides the default connection of ArcGIS Pro to **YouthPopulation.gdb** and other objects of the project, it's sometimes useful to connect to another folder for easy access. The **MaricopaCounty** folder (**Chapter4\Data\MaricopaCounty**) is such a folder: It has spatial data that you'll want to access without having to browse to its location. You will create its connection in the **Catalog** pane.

1. Open the **Catalog** pane.

2. Expand **Folders** and expand **YouthPopulation**.

   Here, you see the contents of the **YouthPopulation** folder, including **YouthPopulation.gdb**.

3. Right-click **Folders** and click **Add Folder Connection**.

4. Browse to **..\EsriPress\GISTforPro\Chapter4\Data**, click **MaricopaCounty** so that it appears in the **Name** field, and click **OK**.

   ArcGIS Pro adds the **MaricopaCounty** folder connection.

5. In the **Catalog** pane, expand **Folders** and the **MaricopaCounty** folder.

You'll import spatial data from the **MaricopaCounty** folder in the next section.

Convert a shapefile to a feature class

A shapefile is a spatial data format for a single point, line, or polygon layer. Although no longer Esri's preferred spatial data format, shapefiles are common and still used by many spatial data suppliers. Because a shapefile doesn't support advanced capabilities, it should be converted to a feature class and stored in a geodatabase.

1. On the **Analysis** tab, in the **Geoprocessing** group, click **Tools**.

2. In the **Geoprocessing** pane, search for and open the **Export Features** tool.

   This tool converts your shapefile to a feature class.

3. For **Input Features**, click the **Browse** button, expand **Folders**, click **MaricopaCounty**, click **Municipalities.shp**, and click **OK**.

4. For **Output Feature Class**, click the **Browse** button, expand **Databases**, click **YouthPopulation.gdb**, and set the **Name** as **Cities**. Click **OK**.

   When you click the **Output Feature Class** field, the full file path appears. It's stored in the default file geodatabase, **YouthPopulation.gdb**, and will be added to the map.

5. Click **Run**.

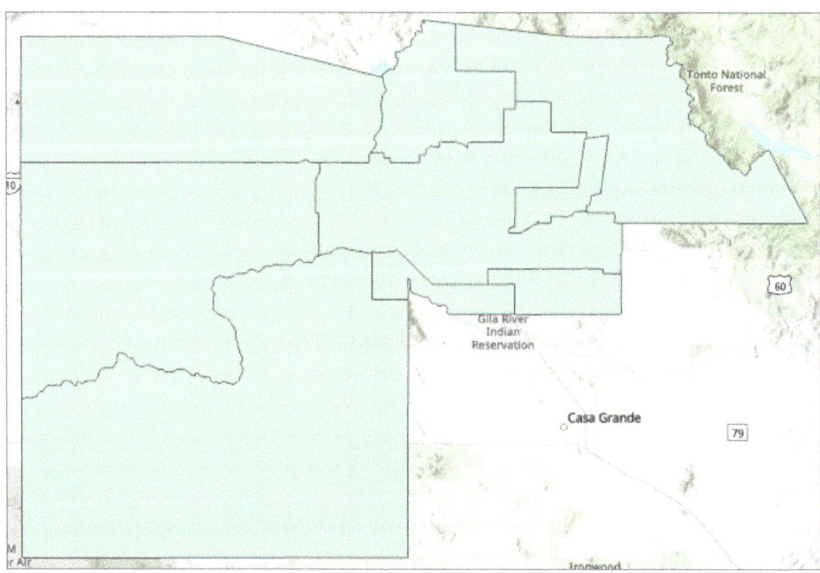

6. Symbolize **Cities** using **Black Outline (2 pts)**. Change the **Outline color** to purple.

> **YOUR TURN**
>
> Repeat the process you just learned. Use the **Export Features** tool, browse to the **MaricopaCounty.shp** and **Tracts.shp** files in the **MaricopaCounty** folder, and create the feature classes **MaricopaCounty** and **Tracts**. Save the output to **YouthPopulation.gdb**. Symbolize each layer with outlines but no fill, giving **MaricopaCounty** the thickest outline and **Tracts** the thinnest. Choose different colors for the outlines. Save your project.

Import a data table into a file geodatabase

Next, you'll import a data table from the American Community Survey (conducted by the US Census Bureau), located in the **MaricopaCounty** folder. The table is in a typical format for data sharing—comma-separated values, or .csv. The vertical columns have attribute names, describing the data in that column, and each horizontal row represents a census tract.

Tip: If you want to import a Microsoft Excel workbook, you can save it as a .csv file and import it.

```
ID,PopTotal,PopUnder5,Pop5To9,Pop10To14,Pop15To19
4013010101,5062,150,105,100,165
4013010102,5127,9,76,57,141
4013030401,5245,0,76,114,241
4013030402,4541,100,98,188,164
4013040502,5304,351,95,115,236
4013040506,5559,0,0,0,0
```

1. Search for and open the **Export Table** tool. Apply the following parameters:
   - **Input Table**: MaricopaCounty > PopYouth.csv
   - **Output Table**: **PopYouth**

2. Click **Run**.

   ArcGIS Pro adds the table to **YouthPopulation.gdb** and the **Contents** pane.

Use database utilities in the Catalog pane

You can create, copy, rename, and delete file geodatabases and anything else in the **Catalog** pane. Suppose you want to make a copy of your data as a backup or to share. Next, you'll create a new file geodatabase, copy spatial data to it, and rename the feature classes and table. Finally, you'll delete a feature class and the file geodatabase.

1. In the **Catalog** pane, right-click **Databases** and click **New File Geodatabase**. For **Name**, type **MaricopaTracts** and click **Save**.

   In **Databases**, **MaricopaTracts.gdb** is added below the default **YouthPopulation.gdb**.

2. In **Databases**, expand **YouthPopulation.gdb**, right-click the **Tracts** feature class, and click **Copy**.

3. Right-click **MaricopaTracts.gdb**, click **Paste**, and expand **MaricopaTracts.gdb**.

   A copy of the **Tracts** feature class has been added.

4. Copy the **PopYouth** table from **YouthPopulation.gdb** to **MaricopaTracts.gdb**.

5. Right-click **MaricopaTracts.gdb** and click **Refresh**.

   You will often need to refresh your geodatabase to see new files you have added or other changes you make.

6. In **MaricopaTracts.gdb**, right-click **PopYouth** and click **Rename**. Rename it **TractsPopYouth**.

   Now you have a stand-alone copy of the **Tracts** feature class and a data table that has been renamed in the **MaricopaTracts** file geodatabase.

   Suppose that you don't want to include the **Tracts** feature class. Next, you'll delete it from **MaricopaTracts.gdb**.

7. In **MaricopaTracts.gdb**, right-click **Tracts**, click **Delete**, and click **Yes**.

   Deleting tables and feature classes from a file geodatabase is permanent. In contrast, removing a layer from the **Contents** pane removes it only from the map and leaves the feature class in a file geodatabase. It can be added to a map and the **Contents** pane again.

Suppose that you no longer want to keep **MaricopaTracts.gdb** or anything inside it. Next, you'll delete it.

8. Right-click **MaricopaTracts.gdb**, click **Delete**, and click **Yes**.

9. Save but do not close your project.

Tutorial 4-2: Modifying attribute tables

Much of what gets processed or displayed in a GIS depends on attributes—columns of data in tables. To get tables in the desired form, you must know how to delete, create, and modify these attributes.

Delete unneeded columns

Many feature classes have extra or unnecessary attributes that you can delete.

1. On the **Project** tab, click **Save Project As** and save your project as **Tutorial 4-2YourName.aprx**.

2. In the **Contents** pane, right-click **Tracts** and click **Zoom To Layer**.

3. Create a bookmark named **Maricopa County**.

4. Turn off both basemaps in the **Contents** pane.

5. In the **Contents** pane, right-click **Tracts** and click **Data Design** > **Fields**.

    This view allows you to create and modify fields in a table. You cannot delete the primary key of the table, **OBJECTID**, or the **Shape**, **Shape_Length**, and **Shape_Area** fields. These fields are essential for a feature class. The text is dimmed, meaning that you can't modify them. **GEOID** will be used later in this chapter, but all other fields are candidates for deletion for the purposes of this chapter.

6. While pressing **Ctrl**, select all rows except **OBJECTID**, **Shape**, **GEOID**, **Shape_Length**, and **Shape_Area**, right-click inside the group, and click **Delete**.

    ArcGIS Pro crosses out the fields but doesn't delete them until you click **Save**. Review selections for deletion before saving. If you decide to keep any crossed-out fields, you can right-click a field and click **Restore**.

7. On the **Fields** tab, in the **Manage Edits** group, click **Save**.

8. Close the **Fields** view. Open the **Tracts** attribute table.

Only the needed attributes remain.

	OBJECTID *	Shape *	GEOID	Shape_Length	Shape_Area
1	1	Polygon	04013618300	0.05062	0.000137
2	2	Polygon	04013116605	0.063523	0.00025
3	3	Polygon	04013116606	0.063491	0.00025
4	4	Polygon	04013811200	0.065178	0.000262
5	5	Polygon	04013216848	0.222358	0.001301
6	6	Polygon	04013216851	0.347328	0.005451
7	7	Polygon	04013613000	0.189639	0.000976

YOUR TURN

Delete all fields from the **Cities** layer except **GEOID**, **NAME**, and the essential fields that are dimmed. While still in the **Fields** view, change the alias for the **NAME** field to **City** and click **Save**. Open the **Cities** attribute table to confirm the updated fields and **City** alias.

Add a field and populate it using the Calculate Field tool

In general, to make a map using US Census Bureau or American Community Survey data, you must download a table with the census data attributes you are interested in from the Census Bureau website (see chapter 5). You should also download the geometry of corresponding census areas (for example, the shapefile for census tracts) from the Census Bureau website. You must then join the data table to the shapefile attribute table based on the geocode field that each file contains. This approach is necessary because there are many thousands of census variables, and it's impractical to supply them all in a single shapefile or feature class; the shapefile would be extremely large and unwieldy.

In this section, you'll prepare a data table (**PopYouth**) so that it can be joined to the attribute table of the **Tracts** feature class on a one-to-one basis. Both tables have 916 rows, and each row or record in the **PopYouth** table has one matching record in **Tracts**. Joining two one-to-one tables increases the number of columns while leaving the number of rows the same. Joining tables is common in GIS and in databases especially. Joining two tables requires each table to have an attribute with matching values stored with the same data type (for example, numeric or text). For the **PopYouth** table, the matching values are census tract geocodes in the **ID** column, whereas for the **Tracts** attribute table, the values are the same geocodes in the **GEOID** column.

The geocodes in **PopYouth** and **Tracts** could be joined, except that **GEOID** in **Tracts** has a text data type and **ID** in **PopYouth** has a numeric data type. Generally, this will be the case: The matching geocodes will have different data types because of the process of downloading and processing data tables. If you open the **Tracts** attribute table, you'll see that the text **GEOID** values for **Tracts** have leading zeros—for example, 04013010101. In contrast, **PopYouth** has numeric values without the leading zeros—for example, 4013010101. The remedy is to create a numeric field in **Tracts**—say, GEOIDNum (to indicate that the data type is numeric)—and copy the data from the **GEOID** field to the new field. The leading zeros are then dropped, and values and data formats will match.

1. With the **Tracts** attribute table open, click its **Options** button in the upper right (three lines) and click **Fields View**.

2. At the bottom of the **Fields** view, click **Click here to add a new field**.

3. For the **Field name**, type **GEOIDNum**, and for **Data Type**, select **Double**.

4. On the **Fields** tab, click **Save**.

5. Close the **Fields** view.

    The values for **GEOIDNum** are <Null>, which signifies that the data cells are empty and no data has been added yet.

6. In the **Tracts** attribute table, right-click **GEOIDNum** and click **Calculate Field**.

7. In the **Calculate Field** pane, double-click **GEOID** in the **Fields** panel to create the expression `GEOIDNum = !GEOID!`.

    The expression's syntax is from the Python programming language, requiring attribute names on the right to be identified using exclamation point delimiters.

    > *Important Note:* The periods used at the end of expressions in the text are placed in the context of a sentence and are not part of the programming code.

8. Click **OK**.

The new attribute, **GEOIDNum**, has **GEOID** values transformed to numeric format without the leading zeros.

OBJECTID *	Shape *	GEOID	Shape_Length	Shape_Area	GEOIDNum
1	1	Polygon	04013618300	0.05062	0.000137	4013618300
2	2	Polygon	04013116605	0.063523	0.00025	4013116605
3	3	Polygon	04013116606	0.063491	0.00025	4013116606
4	4	Polygon	04013811200	0.065178	0.000262	4013811200
5	5	Polygon	04013216848	0.222358	0.001301	4013216848

9. Close the **Tracts** table.

Join a data table to a feature class attribute table

Joining tables is an important and frequent process in databases. With the work completed in the previous section, you can now join tables.

1. In the **Contents** pane, right-click **Tracts** and click **Joins and Relates** > **Add Join**.

   The **Add Join** tool automatically and correctly enters **Tracts** and **PopYouth** tables. The **PopYouth** attributes will be added to the **Tracts** attribute table.

2. In the **Add Join** tool, apply the following settings:
   - **Input Field**: GEOIDNum (Ignore the warning, which states that the join field is not indexed. Indexing [not covered in this tutorial] would speed the join process if the number of rows to be joined was large.)
   - **Join Field**: ID

3. Click **Validate Join**.

   The join is valid.

4. Click **OK**.

5. Open the **Tracts** attribute table.

   All 916 rows of **PopYouth** attributes have been joined to the **Tracts** feature class.

6. Open the **Fields** view of the table.

   Tip: If you do not see the joined attributes, click **Refresh** on the **Fields** tab.

The joined fields have **Field Name** values that include the joined table's name as a prefix—for example, **PopYouth.PopTotal**.

7. Close all open tables.

Export a feature class to make a join permanent

The join you created in the previous section is not a permanent table but is created on the fly each time you start ArcGIS Pro to use the join. This section creates a new feature class based on the joined attributes.

1. In the **Contents** pane, right-click **Tracts** and click **Data** > **Export Features**.

2. In the **Export Features** tool, change the **Output Feature Class** name to **MaricopaTracts**. Click **OK** to run the tool.

3. Open the **Fields** view for **MaricopaTracts**.

   The **PopYouth** table prefixes of joined fields is dropped because all attributes are now permanent as stored attributes of the new feature class.

4. Remove **Tracts** from the **Contents** pane.

Calculate the sum of fields

One purpose of the join you just performed is to map the youth population under the age of 20 in Maricopa County. To make this map, you first need to add the four component attributes of youth populations in smaller ranges making up youths under 20.

1. If not already open, open the **Fields** view of the **MaricopaTracts** feature layer.

2. Add a new field with the name **PopYouthUnder20**. Save and close the **Fields** view.

3. Open the **MaricopaTracts** attribute table, right-click **PopYouthUnder20**, and click **Calculate Field**.

4. In the text box after `PopYouthUnder20 =`, create the expression as follows, double-clicking the corresponding fields in the **Fields** list and clicking the plus sign:

   `!PopUnder5! + !Pop5To9! + !Pop10To14! + !Pop15To19!`

5. Click **Apply** and click **OK**. The **PopYouthUnder20** field is populated with the sums.

PopYouthUnder20
755
2750
2729
583
1396

6. Leave the **MaricopaTracts** attribute table open.

Calculate the percentage of total population under 20 years old

The percentage of the total population that are youths can be found by using the expression `100*PopYouthUnder20/PopTotal`. If you sort **PopTotal** in ascending order, you'll notice that four census tracts have zero population. Because dividing by zero is undefined, it's good practice not to include the four rows with zeros as divisors. In this case, you'll select only rows with **PopTotal** greater than zero. The **Calculate Field** tool processes only selected rows, if any are selected, ignoring unselected rows.

1. Add a new field named **PercentPopYouthUnder20**.

2. On the **Map** tab, click **Select By Attributes**. For **Input Rows**, select **MaricopaTracts**. Create the expression **Where PopTotal is greater than 0** and click **OK**.

3. In the **MaricopaTracts** attribute table, calculate your new field using the following expression:

   `PercentPopYouthUnder20 = 100 * !PopYouthUnder20! / !PopTotal!`

4. Click the **PopTotal** field and click **Sort Ascending**.

5. Because of the selection you made, the **PercentPopYouthUnder20** field was calculated without the census tracts that have zero total population.

PopYouthUnder20	PercentPopYouthUnder20
0	<Null>
0	<Null>
0	<Null>
0	<Null>
0	0
10	40
70	66
34	28

6. Clear the selection and close the table.

YOUR TURN

Symbolize **MaricopaTracts** with **Graduated Colors** using the **PercentPopYouthUnder20** field. Use nine quantiles for the method and choose a green color scheme. In the **Contents** pane, ensure the **Drawing Order** is as follows: **MaricopaCounty** > **Cities** > **MaricopaTracts**. You can now see areas with high percentages of youths where city policymakers may want to consider improving youth services or putting new youth services.

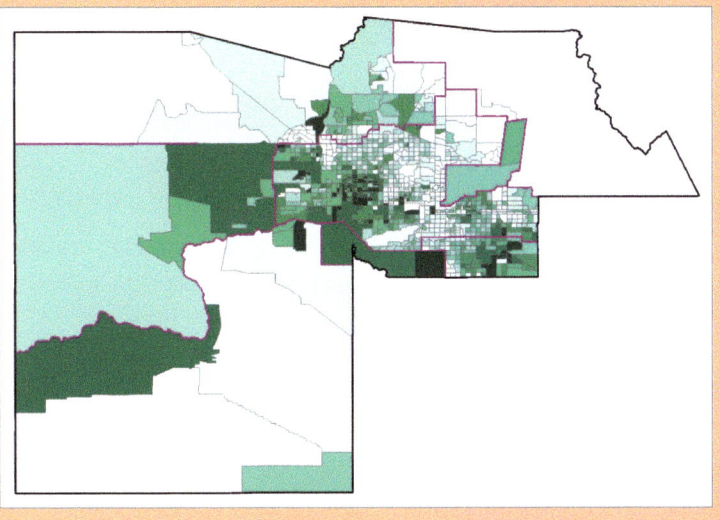

Extract substring fields and concatenate string fields

In this section, you'll learn sophisticated techniques for working with text fields, extracting parts of text strings, and reassembling them into a new text field.

1. In the **MaricopaTracts** attribute table, sort **GEOID** in ascending order.

   The first value is **04013010101**. This unique identifier for a census tract is composed of three parts:
   - 04 is the American National Standards Institute (ANSI) code for Arizona.
   - 04013 is the ANSI code for Maricopa County in Arizona.
   - 04013010101 is the ANSI identifier for a census tract in Maricopa County, Arizona.

   The census tract component, 010101, is further broken down into the first four digits (0101) and an optional two-digit suffix (01) with an implied decimal point between the number and suffix. Assuming that users know that the data is for Maricopa County, the corresponding census tract name is Census Tract 0101.01. Using **GEOID** as the input, you'll compute this representation of tracts.

2. In the **MaricopaTracts** attribute table, create three new text fields. Set the length using the cell at the far right of the **Fields** table. Use these specifications for the fields:
   - **Field Name:** TractNumber, **Data Type:** Text, **Length:** 4
   - **Field Name:** TractSuffix, **Data Type:** Text, **Length:** 2
   - **Field Name:** TractName, **Data Type:** Text, **Length:** 20

3. Calculate the **TractNumber** field with the Python expression: `!GEOID![5:9]`.

   To understand the Python language syntax of this expression, consider the case of tract 04013123456. The characters stored (digits, in this case) of any field are indexed 0, 1, 2, ..., from left to right in Python, so the indices of the 04 for Arizona are 0 and 1. The notation `[0:2]` extracts the indexed first and second digits and up to but not including the third digit, 0. So 04 is extracted. For the case of extracting **TractNumber** and the example of 04013123456 and `!GEOID![5:9]`, start counting from zero and find the fifth digit (1), where the extraction begins, and continue until you reach the ninth digit (5), which is not included. So `[5:9]` extracts 1234. If this notation is confusing, you can always use trial and error, calculating a field until you get it right.

4. Calculate the **TractSuffix** field with the expression `!GEOID![9:11]`.

This expression extracts the ninth digit up to but not including the 11th character and produces the value 01.

5. Calculate **TractName** with the following expression:

   ```
   "Census Tract " + !TractNumber! + "." + !TractSuffix!
   ```

   Tip: There is a space between `Census Tract` and the closing double quotation mark. The plus sign concatenates or combines two text values into one. "Census Tract " and "." in double quotation marks are constants, the same for every row of data. `!TractNumber!` and `!TractSuffix!` with exclamation point delimiters denote field names, which vary by row.

6. Sort in ascending order by **TractName**.

TractNumber	TractSuffix	TractName
0101	01	Census Tract 0101.01
0101	02	Census Tract 0101.02
0304	01	Census Tract 0304.01
0304	02	Census Tract 0304.02
0405	02	Census Tract 0405.02

Tutorial 4-3: Carrying out attribute queries

One of the major innovations of GIS technology is linking tabular data to the spatial features in feature classes. This linkage allows you to symbolize maps using the attribute values found in tables and enables you to find spatial features of interest using attribute data. An attribute query selects attribute data rows and spatial features based on attribute values.

Attribute queries are based on SQL, the de facto standard query language of databases and many apps, including ArcGIS Pro. This tutorial introduces the query criteria of SQL. For further study and to learn full SQL commands, you can find interactive tutorials on the internet. The next two tutorials use crime data and queries commonly made when analyzing crime patterns. The example queries ask some of the standard query questions of who, what, where, when, and how.

A simple SQL criterion has the following form:

```
attribute name <logical operator> attribute value.
```

The attribute name can be any attribute column heading or field name in an attribute table. Several logical operators are available, including the familiar ones, such as =, >, >=, <, and <=. The attribute value specifies what you are looking for. For example, the following simple criterion selects all crimes that are robberies where **Robbery** is a value of the **Crime** attribute:

```
Crime = 'Robbery'.
```

Text values stored in attributes, such as **Robbery**, must be enclosed in single or double quotation marks. Numeric values, stored in numeric fields, do not need quotation marks.

You can use queries to select points in a feature class. A point that is selected appears highlighted in the selection color, in the attribute table and on the map, and all points of the layer remain visible.

Compound SQL criteria are made up of two or more simple criteria connected with either an AND or an OR connector. AND means that both connected simple criteria must be true for corresponding features (records) to be selected. For example, for SQL to select features for the compound query

```
DateOccur >= date '2015-08-01' AND
DateOccur <= date '2015-08-28',
```

the **DateOccur** values must be greater than or equal to August 1, 2015, and less than or equal to August 28, 2015. SQL will select a feature with the date August 5, 2015, for example, but it will exclude August 31, 2015. The date values in the preceding criteria are in the format that ArcGIS Pro requires for use in queries. You don't have to remember such formats because the ArcGIS Pro query builder helps you get the values (and in the right format) you need in queries, as you will see in this tutorial.

The OR connector can also be used to select criteria, such as crime types in the **Crime** attribute. This connector should be used if you want records that satisfy more than one criterion—for example, more than one type of crime. To ask for this kind of data in conversation, you might say, "I want data on all homicides and aggravated assaults," but this statement is ambiguous: It may mean that you want all data in which there was both a homicide and an aggravated assault in a single incident or that you want all data in which either a homicide or an aggravated assault occurred. SQL avoids this problem by requiring specific language; to get data in which either crime occurred, the expression is "homicides OR aggravated assaults." So, the proper SQL criteria in this case are

```
Crime = 'Homicide' OR Crime = 'Aggravated Assault'.
```

Despite occasional exceptions to this rule, you will almost always need to place an OR criterion in parentheses (as shown) if the criteria are combined with other criteria, such as a date range:

```
DateOccur >= date '2015-08-01' AND
DateOccur <= date '2015-08-28' AND
(Crime = 'Homicide' OR
Crime = 'Aggravated Assault').
```

The use of parentheses (just as with algebraic expressions) is essential because logical expressions are run one pair at a time for simple expressions, generally working from left to right but with certain logical operators going first. For example, SQL runs AND comparisons before OR comparisons regardless of order, which can result in incorrect information unless you use parentheses to control the order of running. SQL runs comparisons in parentheses first. For example, if you leave the parentheses out of the previous query, you'd retrieve all homicides for the specified period in August 2015 but also all aggravated assaults in the database regardless of the date. Verifying input data and the output results will give you experience and guidance in making compound queries.

Crime analysts use three primary kinds of attribute queries, which can be combined in the following ways:

- The most fundamental attribute query is by crime type and data interval (the *what* and *when*). Such queries often combine several crime types with the use of logical operators.
- The second type of attribute query adds criteria, such as the time of day or day of the week—for example, weekday versus weekend crime (a refinement of *when*). Clearly, crime patterns can differ by the time of day or day of the week.
- The third type of attribute query adds criteria based on the attributes of the people (the *who*) or the objects (the *what*), such as vehicles, involved in a crime and can include modus operandi attributes (the *how*), such as "Enters through an open window," if available.

View crime incidents

1. Open **Tutorial4-3** in **Chapter4\Tutorials** and save it as **Tutorial4-3YourName**.

The map of Pittsburgh has the feature layer **Crime Offenses**, which has all criminal offenses committed for June through August 2015 recorded by the Pittsburgh Bureau of Police, plus **Streets** and **Neighborhoods**. Symbology for **Crime Offenses** is **Unique Values** based on the **Crime** attribute, so the crime data is comprehensive, with all 26 crime types recognized by the FBI. It's not possible to analyze crime patterns in the current map with all crime types plotted, so you'll examine one or a few crime types at a time to identify spatial patterns. The crime data, although real, has been modified to protect privacy, including changing the year of crimes, adding small random numbers to house numbers in incident addresses, and using fictitious names and addresses of persons. Also, Pittsburgh is a relatively low-crime city, and crime data plotted for any city would likely look like the crime data in this tutorial: It seems like a lot of crime.

2. In the **Contents** pane, right-click **Crime Offenses** and click **Symbology**.

3. In the **Symbology** pane, click **More** > **Show count**.

    The count data provides the frequency of crimes by crime. You'll make a map for **Burglary** (frequency 846) and **Robbery** (frequency 412).

4. Close the **Symbology** pane.

Create a date-range selection query

Queries for event locations, such as crimes, almost always use date-range criteria. Once you select the features, you can perform additional processing of selected outputs.

1. On the **Map** tab, in the **Selection** group, click **Select By Attributes**.
    - Create the expression **Where DateOccur is on or after 7/1/2015 12:00:00 AM**, selecting the date from the list.
    - Click the **Verify** button (green check mark). The **Verify** button ensures that there are no syntax errors but doesn't check for logic errors.

- Click **Add Clause** and add to the query so the full expression reads, **Where DateOccur is on or after 7/1/2015 12:00:00 AM And DateOccur is on or before 7/31/15 12:00:00 AM**.

The data you're analyzing has time of crime occurrences in a separate attribute, **TimeOccur**. So the 12:00:00 AM that is part of the **DateOccur** attribute is not valid. You can ignore it.

2. Click **Apply**.

   All crime offenses remain on the map, but the points for July 2015 are now displayed in the selection color.

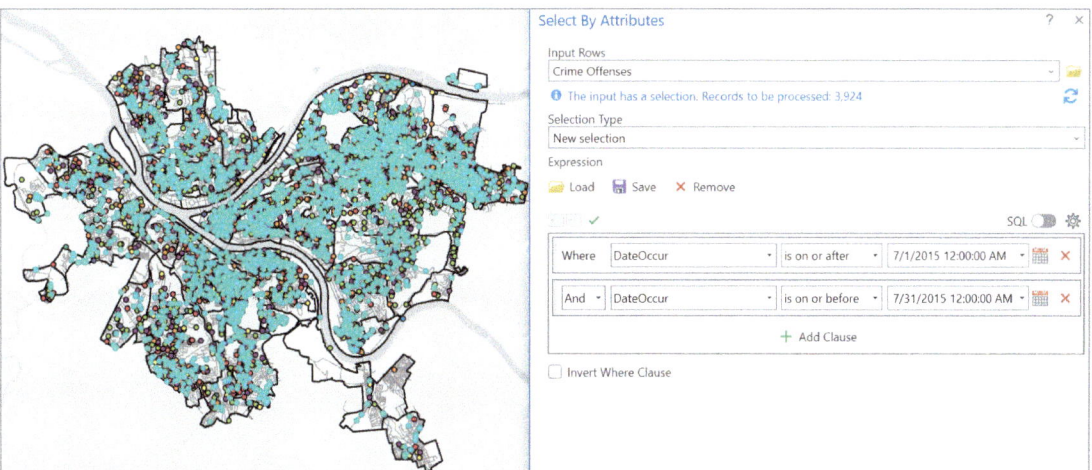

3. In the **Select By Attributes** tool, under **Expression**, click **Save**, and save the expression as **qryDateRange** in **Chapter4\Data**.

   It's not required to save queries, but in the next section, you'll reuse or reload this query for a definition query. The "qry" prefix is a standard prefix for database queries.

4. Close the **Select By Attributes** tool and open the **Crime Offenses** attribute table. In the lower left of the table, click the **Show Selected Records** button.

   Notice that 3,924 out of 11,500 features are selected.

5. Close the attribute table and clear the selection.

Reuse a saved query to create a definition query

1. In the **Contents** pane, right-click **Crime Offenses**, click **Properties**, and click **Definition Query**. Click the arrow to open the **New Definition Query** menu, click **Add definition queries from files**, and double-click **qryDateRange.exp**.

   The saved expression reloads. If you had not saved the query in the previous section, you would have had to recreate it here.

2. Click **OK**.

3. Open the **Crime Offenses** attribute table, and in the lower left of the table, click **Load All**. Verify that only the 3,924 July 2015 crimes remain.

4. Close the table.

Query a subset of crime types using OR connectors and parentheses

Next, you'll add clauses to query burglaries and robberies.

1. Right-click **Crime Offenses** and click **Properties** > **Definition Query** > **Edit** to open the **Definition Query** property sheet.

   Your two date clauses connected with AND are still there. You'll add two clauses, one for burglaries and one for robberies, connected by ORs.

2. Click **Add Clause** and add the clause **And Crime Is Equal to Burglary**.

3. Click **Add Clause** and add the clause **Or Crime Is Equal to Robbery**.

   If you ran this query without parentheses around the OR clauses, you'd get disorderly conduct features in July 2015 and then all vagrancy and vandalism in the dataset (not just in July).

4. Turn on the **SQL** toggle key.

   This step shows you the actual SQL criteria that the query builder built, which is mostly self-explanatory.

5. Type an opening parenthesis just before the first "Crime" and a closing parenthesis at the very end, like the following:

    … And (Crime = 'Burglary' Or Crime = 'Robbery')

   SQL first evaluates the expression inside the parentheses, which is to select all burglary or robbery crimes. SQL then works from left to right to select all records in July for the two crime types.

   ```
   DateOccur >= timestamp '2015-07-01 00:00:00' And
     DateOccur <= timestamp '2015-07-31 00:00:00' And
     (Crime = 'Burglary' Or Crime = 'Robbery')
   ```

6. Click **Apply** and click **OK**. Open the **Crime Offenses** attribute table and verify that the 444 remaining features are for the two crimes of interest in July.

   There are 309 burglaries and 135 robberies mapped.

7. Symbolize **Robbery** with a dark-red color.

   Generally, violent crimes, such as robberies, are clustered in specific areas, whereas burglaries are widespread. You can see some evidence of these patterns in the map.

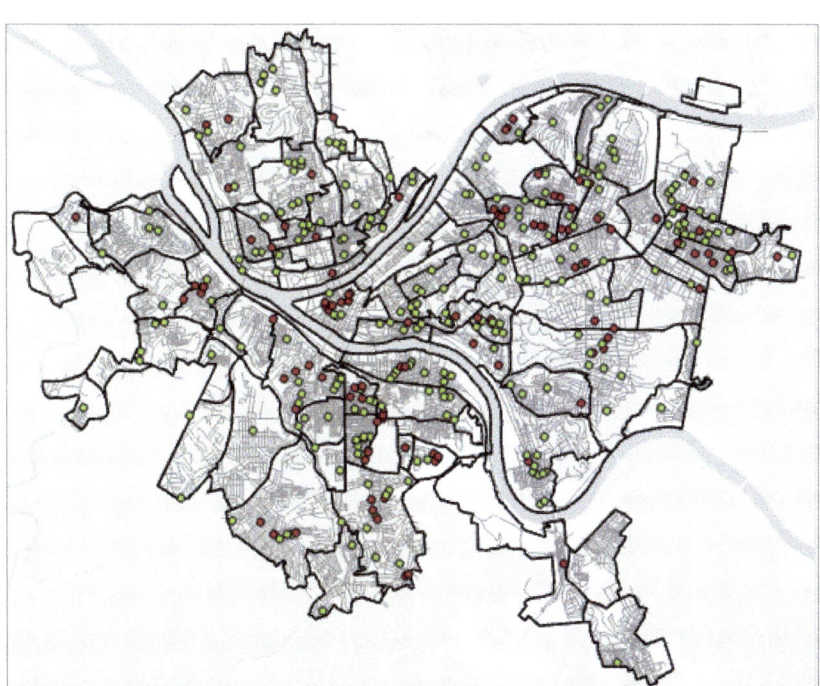

8. Open **Definition Query** for **Crime Offenses**. In the list of **Definition Queries**, point to the **qryDateRange** expression, and click the **Clear Active Query** button (green check mark).

Now you're back to having all original features displayed.

> **YOUR TURN**
>
> Create a definition query for burglaries in August 2015. Try editing the current definition query. You can use the **SQL** view to remove the parentheses, which are no longer needed but cause no harm. Open the attribute table to verify the results. There should be 273 burglary records. Symbolize them with a bright-red color.

Query the day-of-week range

Next, you'll use the attribute **DayOfWeek** in **Crime Offenses**, which identifies the day of the week for burglaries in August 2015. You'll build a selection query for burglaries over the weekend and obtain the number of weekend burglaries in August. You'll switch the selection and obtain the number of weekday burglaries. Finally, you'll obtain the average number of burglaries per day on a weekday versus a weekend day. Which do you think will be higher?

1. Click **Select By Attributes** and clear the existing query.

2. Create a new query with the following expression: **Where DayOfWeek is equal to Saturday or DayOfWeek Is Equal to Sunday**.

3. Click **OK**.

4. Open the **Crime Offenses** attribute table.

   Notice that 84 of the 273 burglaries are selected.

5. In the attribute table, click **Switch** to switch the selection.

   Notice that 189 of the 273 features are weekday burglaries.

   August 2015 had 10 weekend days and 21 weekdays. Doing the math, an average of 8.4 burglaries occurred per weekend day and 9.0 occurred each weekday. Burglaries often occur when residents are away from home, so one expects weekdays to have more burglaries than weekends, although the difference is small in this case.

6. Clear the selection.

    The definition query is still set to August burglaries, so 273 burglaries are mapped with none selected.

Query person attributes

Suppose an informant told a police officer that a burglary was committed by a male in his 30s who lives on Warrington Avenue, and the officer wants to see whether a person with those characteristics has been arrested for burglary. The query will need several clauses all connected by AND because the person sought must have all the provided characteristics. Also, you must search for a part of a text value—namely the street name (Warrington) from a field with a full address (such as 123 Warrington Ave.).

1. Start a new **Select By Attributes** query for **Crime Offenses**. Ensure that all the previous clauses are cleared.

    This time, you'll keep applying the query as you build it and watch as the query results are narrowed. To get the arrests, you'll start by excluding all records in which there was no arrest, which you'll find by querying for the arrested person's last name as not being null. Null means no value entered, which indicates no arrest.

2. Add the clause **Where ArrLName is not null**, click **Apply**, and click **Show Selected Records**.

    Forty arrests were made for burglaries committed in August 2015, all of them now visible and in the selection color.

3. Add and apply the clause **And ArrSex is equal to M**.

    Now there are 34 records selected.

4. Add and run two AND clauses for **ArrAge is greater than or equal to 30** and **ArrAge is less than 40**.

    You're down to nine males in their 30s who were arrested for burglary. You can review the records themselves to visually apply the last criterion and street name, but for other cases this method can be inefficient. So, you'll add the last clause for street name.

5. Add and run the AND clause **ArrResid contains the text WARRINGTON**.

   When you check the list of options to complete the query, you'll notice that the street names are in capital letters (for example, **WARRINGTON**). Because SQL is case sensitive for values, you must use all capital letters.

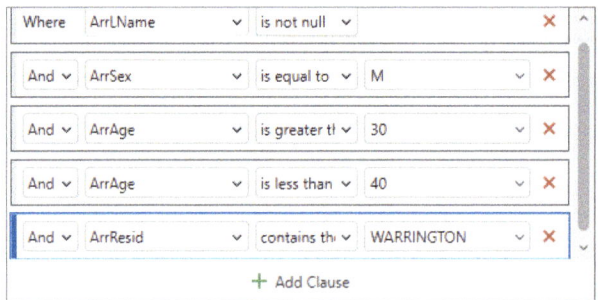

6. Enable the **SQL** view to see the command you just built.

   The SQL logical operator LIKE is used only on text-valued attributes. The value **WARRINGTON** is enclosed in single quotation marks so that SQL learns that it's a text value, and the percent (%) symbol is the wildcard character for SQL that stands for zero, one, or more characters to ignore. The record you'll retrieve has the value **1005 E WARRINGTON AV** for **ArrResid**. The first % in the query value (**'%WARRINGTON%'**) ignores characters at the beginning of the address (**1005 E**), and the last % ignores the characters at the end (**AV**).

7. Click **OK** to review the record in the attribute table.

   You now can identify the person who may have committed the unsolved burglary, John Bond.

8. Clear the selection, close the table, and save your project.

Tutorial 4-4: Aggregating data with spatial joins

In this tutorial, you'll count (aggregate) burglaries by neighborhood and then map the burglaries by neighborhood, giving an aggregate view of those crimes. Aggregation of point data requires a spatial join of burglary points to neighborhood polygons. The spatial join algorithm determines the polygon in which a point lies, enabling the data aggregation.

Build a spatial join

1. Open **Tutorial4-4.aprx** in **Chapter4\Tutorials** and save it as **Tutorial 4-4YourName.aprx**.

   This map is similar to one you used in tutorial 4-3.

2. Search for and open the **Spatial Join** tool and apply these settings:
   - **Target Features**: Neighborhoods
   - **Join Features**: Crime Offenses
   - **Output Feature Class**: **August2015BurglariesByNeighborhood**

3. Click **Run**.

4. Open the new layer's attribute table. Find the **Join_Count** field, which has been added as a new attribute, and notice that it's equal to the number of burglaries in each neighborhood.

5. Close the table.

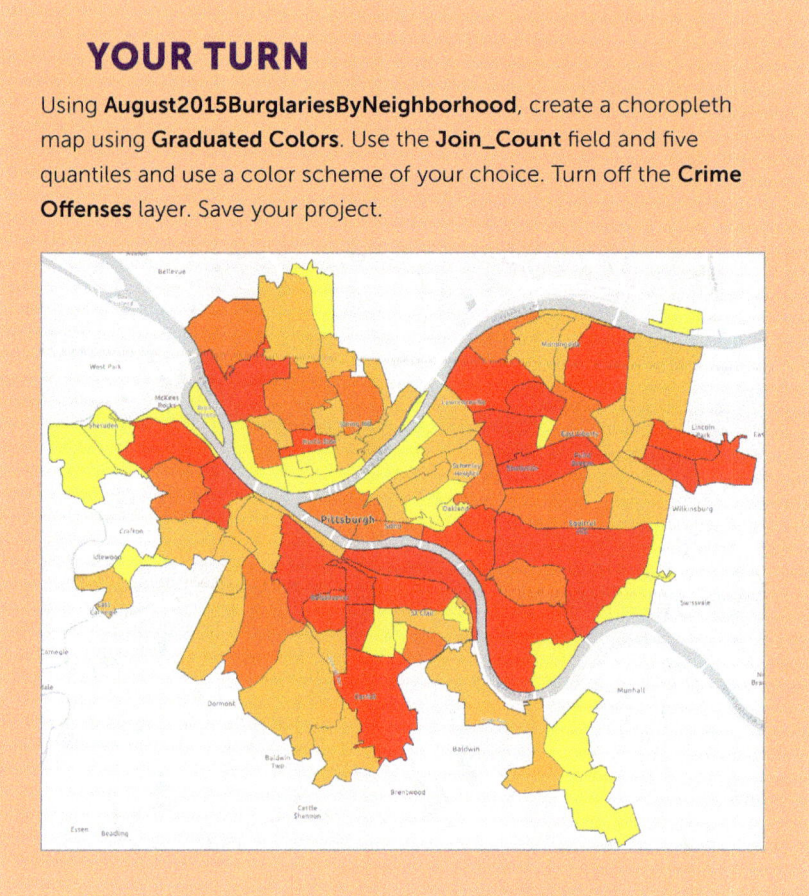

YOUR TURN

Using **August2015BurglariesByNeighborhood**, create a choropleth map using **Graduated Colors**. Use the **Join_Count** field and five quantiles and use a color scheme of your choice. Turn off the **Crime Offenses** layer. Save your project.

Tutorial 4-5: Using central point features for polygons

If you choose graduated symbols for symbology for a polygon layer, ArcGIS Pro creates the central points on the fly and renders them as point features. This choice works well and saves you the work of creating a point feature class from the polygons. However, it's often helpful to make a separate feature class with central points created from polygon features.

Create a central point feature class for polygons

The centroid of a polygon is the arithmetic mean of all points within the polygon. For most polygons, centroids lie within their polygons, but for some, such as a quarter-moon-shaped polygon, centroids lie outside. If you want all center points to lie within their polygons, the remedy in ArcGIS is to use central points instead of centroids. Central points all lie within their polygons. In this tutorial, you'll use the **Feature to Point** tool to create a central point feature class. Although not necessary, for practice you'll first use the **Calculate Geometry Attributes** tool for adding central point coordinate attributes to the feature class.

1. Open **Tutorial4-5.aprx** from **Chapter4 \Tutorials** and save it as **Tutorial 4-5YourName.aprx**.

   This map shows the number of burglaries in Pittsburgh by neighborhood.

2. Search for and open the **Calculate Geometry Attributes tool**, apply these settings, and run the tool:
   - **Input Features**: Burglaries by Neighborhood
   - **Field**: **X**
   - **Property**: Central point x-coordinate
   - **Field**: **Y**
   - **Property**: Central point y-coordinate

3. Open the **Burglaries by Neighborhood** attribute table.

   The tool created fields with the x and y central point coordinates, in state plane feet.

4. Close the attribute table.

Create a point layer

The **Feature to Point** tool creates a point layer of central points for polygons.

1. Search for and open the **Feature to Point tool** and apply these settings:
   - **Input Features**: Burglaries by Neighborhood.
   - **Output Feature Class**: **BurglariesByNeighborhoodPoints**.
   - Check the box for **Inside**.

   This option calculates central points instead of centroids.

2. Run the tool.

   You now have a point layer that you can combine with another layer to indicate some other attribute, such as neighborhoods living in poverty.

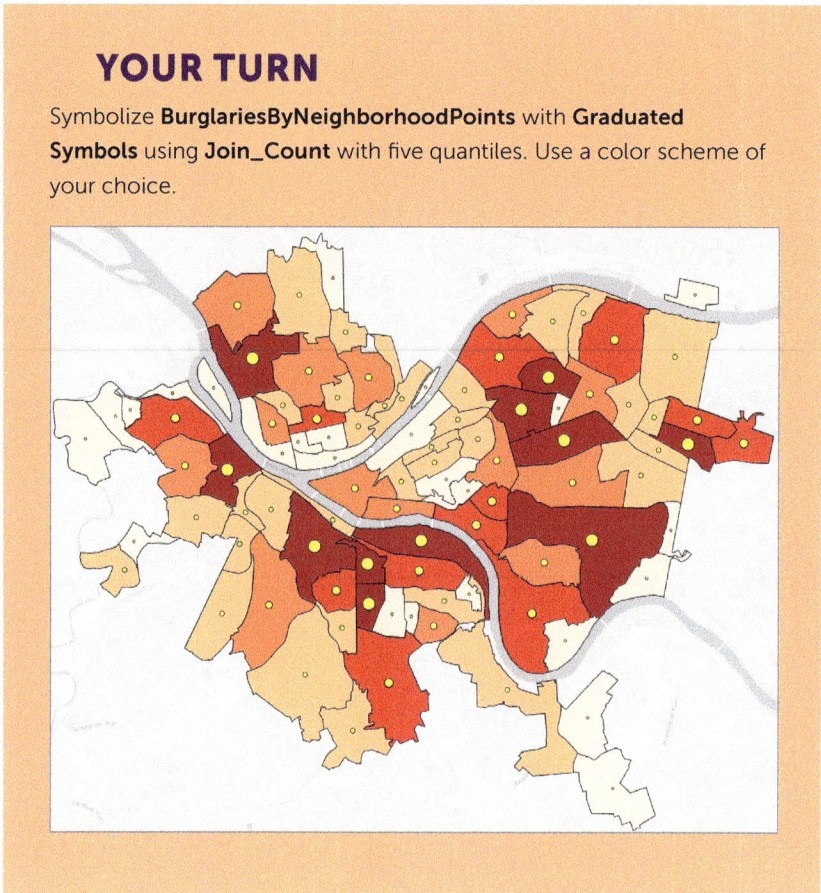

YOUR TURN

Symbolize **BurglariesByNeighborhoodPoints** with **Graduated Symbols** using **Join_Count** with five quantiles. Use a color scheme of your choice.

Tutorial 4-6: Creating a new table for a one-to-many join

Sometimes a data table has a field name that uses a code that, by itself, isn't easily understood. For example, crime data may have an FBI hierarchy code number to represent the crime type. To understand what the numbers mean, you need a code table with the number in one field and the description in a second field. In this tutorial, you'll create the code table and join it to the crime data. This join is called one-to-many, because the same code table record—for example, the 3 for robbery—is joined to many crime incidents, for all robberies.

Create a table

1. Open **Tutorial4-6.aprx** from **Chapter4\Tutorials** and save it as **Tutorial 4-6YourName.aprx**.

2. Search for and open the **Create Table tool**.

3. For **Table Name**, type **UCRHierarchyCode**, and run the tool.

4. Open the **Fields** design view of **UCRHierarchyCode** and add two new fields: **Hierarchy,** with the **Short** data type, and **CrimeType,** with the **Text** data type and length **25**. Save and close the **Fields** view.

5. Open the **UCRHierarchyCode** table and add a new row. For **Hierarchy**, type **3**. For **CrimeType**, type **Robbery**. Press **Enter**.

> **YOUR TURN**
>
> Follow the pattern to complete the rest of the table: **4**, **Aggravated Assault**; **5**, **Burglary**; **6**, **Larceny − Theft**. On the **Edit** tab, click **Save**.

Make a one-to-many join

You'll join your new code table to the map's crime layer to make the crime type easy to understand.

1. Open the attribute table for **PittsburghSeriousCrimesSummer2015** and review the **Hierarchy** field.

   The field is difficult to interpret because it lists only the code, not the crime type.

2. Right-click **PittsburghSeriousCrimesSummer2015** and click **Joins and Relates** > **Add Join**. For **Input Field** and **Join Field**, use the **Hierarchy** field. Click **OK**.

3. View the **PittsburghSeriousCrimesSummer2015** attribute table again to see the joined attributes.

> **YOUR TURN**
>
> Symbolize **PittsburghSeriousCrimesSummer2015** with **Unique Values**. For **Field**, use **CrimeType**. If you want, you can try symbolizing crimes by giving each crime a different symbol. By creating and joining the code table, you can see the legend in the Contents pane is now labeled with crime types. When you finish, save and close your project.

Assignments

This chapter has assignments to complete that you can download with data from ArcGIS Online at links.esri.com/GISTforPro3.4Assignments.

CHAPTER 5

Spatial data

LEARNING GOALS

- Work with world map projections.
- Work with US map projections.
- Set a projected coordinate system.
- Work with vector data formats.
- Work with US Census Bureau spatial and tabular data.
- Explore sources of spatial data.

Introduction

The most important information that maps and GIS provide is location. Where is an object or event positioned? Is the object on a major street with good access? Is there a barrier to accessing it, such as a river? Is it in a certain area? Where are similar things positioned? What or who is near it? These are some of the questions that maps can answer.

Where you sit right now has two unique numbers, latitude and longitude coordinates that pinpoint your location precisely on the surface of the earth. In this chapter, you will learn about latitude and longitude coordinates and their geographic coordinate system.

You will also learn about map projections, making flat maps from the nearly spherical earth. The projection you choose can make a big difference, so this chapter provides some guidelines for choosing a projection.

For the world map in tutorial 5-1, you'll change the geographic coordinate system to a projected coordinate system by setting the map properties.

For the continental-scale map in tutorial 5-2, you'll change the projected coordinate system to a different projected coordinate system by setting the map properties.

In tutorial 5-3, you'll set the projected coordinate system (state plane) for a local-level map by adding a first layer to the map and specifying the display units for the map. You'll add a second layer, which has geographic coordinates, and see how it's projected on the fly to state plane coordinates. Finally, you'll change the geographic coordinate system for a state-level map to use a projected coordinate system by setting the map properties.

Tutorials 5-4 to 5-6 will already have the appropriate projected coordinate systems set for each map.

The key input of GIS is geospatial data—digital data that represents, and can be rendered into, maps on a computer. You may collect some original geospatial data yourself, but more likely, most of your map data will come from external organizations. Who collects aerial imagery, roads, population, or other map data that you will use? The answer is that international organizations and federal, state, and local governments collect the data and will provide it to you at little or no cost, as a download or increasingly as a service from the internet. ArcGIS Living Atlas of the World, available in ArcGIS Online, is one of the foremost sites for obtaining such data. You will explore ArcGIS Living Atlas and both a US federal and a local data source in this chapter.

Tutorial 5-1: Working with world map projections

ArcGIS Pro has more than 5,200 projected coordinate systems (and almost 600 geographic coordinate systems) that use more than 100 map projections from which you may choose. Typically, though, you will need relatively few projections. Geographic coordinate systems use latitude and longitude coordinates for locations on the surface of the earth, whereas projected coordinate systems use a mathematical transformation from an ellipsoid or a sphere to a flat surface and two-dimensional coordinates. Geographic coordinates are angles calculated from the intersection of the prime meridian (which runs north and south through Greenwich, England) and the equator. Longitude, which measures east–west, ranges from 0 degrees to 180 degrees east and the same to the west. Latitude, which measures north–south, ranges from 0 degrees to 90 degrees north and the same to the south. Although you can view 2D maps in geographic coordinates on your flat computer screen, they are greatly distorted because their coordinates are from a sphere. In ArcGIS Pro, you can switch between coordinate systems and map projections on the fly. For world maps, you'll change the geographic coordinate system to a projected coordinate system by setting the map properties.

Examine a world map in geographic coordinates

This section shows you the distortions caused by displaying a map in geographic latitude and longitude coordinates.

1. Open **Tutorial5-1.aprx** from **Chapter5\Tutorials** and save it as **Tutorial 5-1YourName.aprx**.

2. Zoom to the full extent.

The map has significant distortions. For example, the line running across the top and bottom of the map should be points, the North and South Poles. Also, Antarctica and Greenland are far larger proportionately than reality. Although perhaps not evident, the 48 contiguous US states are squashed in the vertical dimension, as are Europe and Asia.

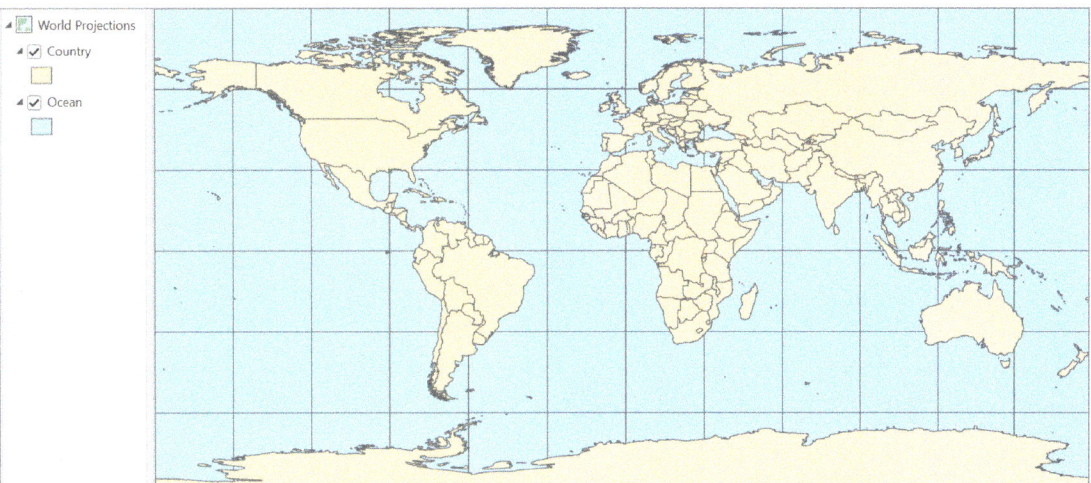

3. Point to the westernmost tip of Africa and read the coordinates on the bottom of the map window (reading approximately 17° W, 20° N).

   The network of lines on the map is called a graticule, and it has 30-degree intervals east–west and north–south.

Project the map on the fly to Hammer-Aitoff (world)

The Hammer-Aitoff projected coordinate system works well on a world map because it is an equal-area projection that preserves area. For example, if you want to map population densities (such as population per square mile), the densities will be correct.

1. In the **Contents** pane, right-click **World Projections** > **Properties**. Click the **Coordinate Systems** tab.

   The current coordinate system is **GCS WGS 1984**, which stands for geographic coordinate system World Geodetic System 1984.

2. Under **XY Coordinate Systems Available**, scroll down, expand **Projected Coordinate System**. Double-click **World** and click **Hammer-Aitoff (world)**.

3. Click **OK** and zoom to the full extent.

The map displayed here has coordinates projected on the fly that are not permanent changes to the feature class.

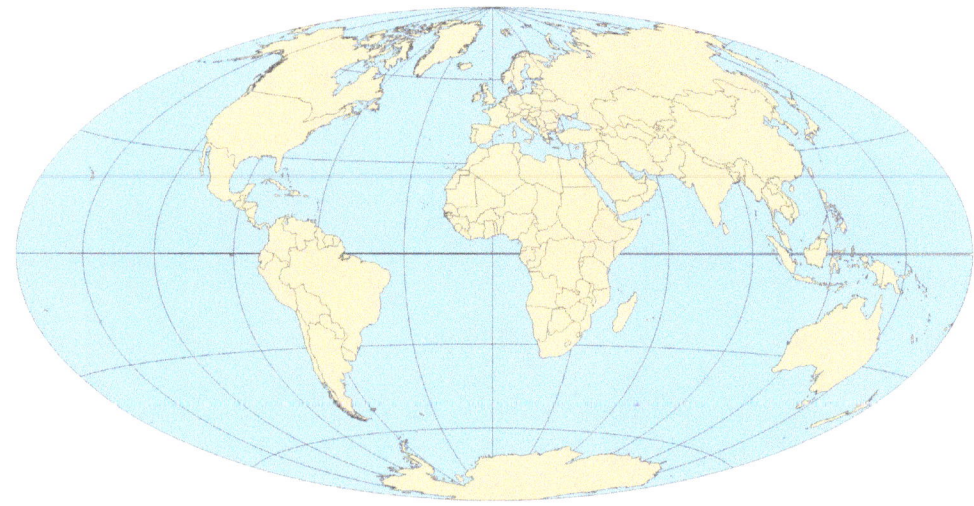

YOUR TURN

Repeat the steps of this section, but this time, select the **Robinson (world)** projection in step 2. The Robinson projection is most accurate at the mid-latitudes in both the Northern and Southern Hemispheres, where most people live, and overall, it is visually appealing. This projected coordinate system minimizes some distortions but still has notable ones, such as the lines at the top and bottom that are really points. As a rule, though, use the Robinson projection for the entire world unless you have a specific need (such as mapping population densities, in which case, use an equal-area projection). Save your project.

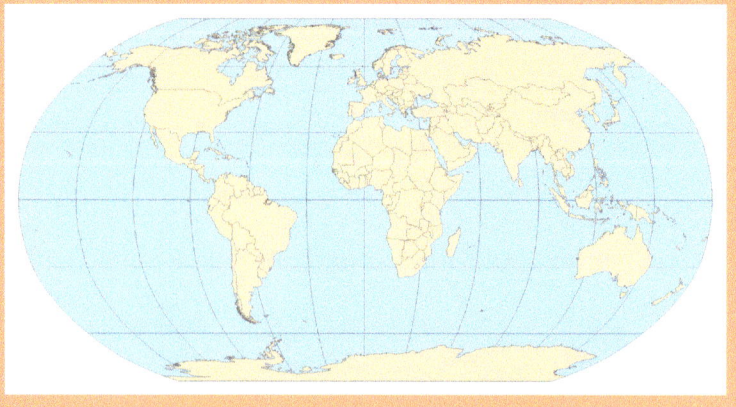

Tutorial 5-2: Working with US map projections

Next, you will gain experience with projections commonly used for maps of the continental United States. You can get accurate areas or accurate shapes and angles, but not both, when using a projected coordinate system. As a rule, use a projection that provides accurate areas (at the price of some shape and direction distortion), such as the Albers equal-area or cylindrical equal-area projections. Albers equal-area is the standard projection of both the US Geological Survey (USGS) and the US Census Bureau for US maps. For a continental-level map, you'll change the projected coordinate system to a different projected coordinate system by setting the map properties.

Set projected coordinate systems for the United States

1. Open **Tutorial5-2.aprx** from **Chapter5\Tutorials** and save it as **Tutorial 5-2YourName.aprx**.

   Initially, the map display is in projected coordinates using **WGS 1984 Web Mercator (auxiliary sphere)**, which is the preferred projected coordinate system for maps to be published in ArcGIS Online.

2. In the **Contents** pane, right-click **US Projections** > **Properties** > **Coordinate Systems**, and scroll up and expand **Projected Coordinate System** > **Continental** > **North America**.

3. Click **USA Contiguous Albers Equal Area Conic** and click **OK**.

4. Zoom in to the 48 contiguous states.

   Your map is now using the Albers equal-area projection.

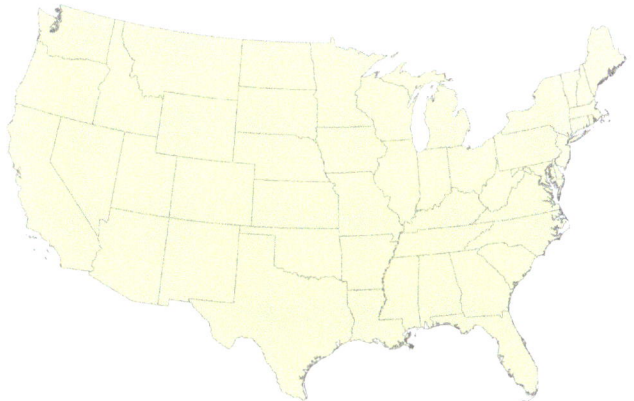

> **YOUR TURN**
>
> Experiment by applying a few other projections to the US map, such as **North America equidistant conic**. If you stay in the same group—**Continental**, **North America**—all the projections look similar. The conclusion is that the larger the part of the world that you need to project (small-scale maps), the more distortion. Much distortion remains at the scale of a continent, but much less so than for the entire world. By the time you get to a part of a state, such as Allegheny County, Pennsylvania, practically no distortion is left, as you will see next. Save your project.

Tutorial 5-3: Setting projected coordinate systems

For medium- and large-scale maps, use localized projected coordinate systems, tuned for the study area, that have little or minimal distortion. For this purpose, collections of projected coordinate systems are divided into zones. You must use a reference map to determine which zone your study area is in and select that projected coordinate system for your map. In this tutorial, you'll set the projected coordinate system (state plane) for a local-level map by adding a first layer to the map, and you'll specify the display units for the map. Then you'll add a second layer that has geographic coordinates and see how that layer is projected on the fly to the state plane coordinates. Finally, you'll change the geographic coordinate system for a state-level map to use a projected coordinate system by setting the map properties.

Look up a zone in the state plane coordinate system

The state plane coordinate system is a set of coordinate systems dividing the 50 US states, Puerto Rico, and the US Virgin Islands, American Samoa, and Guam into 126 numbered zones, each composed of counties, and with its own finely tuned map projection. Used mostly by local government agencies, such as counties, municipalities, and cities, the state plane coordinate system is for large-scale mapping in the United States. The National Geodetic Survey (then known as the US Coast and Geodetic Survey) developed this coordinate system in the 1930s to provide a common reference system for surveyors and mapmakers. Most states use NAD83, which stands for North American Datum of 1983, the datum used to describe the geodetic network in North America. That datum in turn was updated from the original North American Datum of 1927 when satellite geodesy and remote sensing technology became more precise and were made available for civilian applications. The first step in using the state plane coordinate system is to look up the correct zone for your area and, consequently, a specific projected coordinate system tailored to your study area.

1. In a web browser, browse to ArcGIS Living Atlas, at **livingatlas.arcgis.com**.

2. In the search box, type **USA State Plane Zones NAD83**.

3. Choose the web map by Esri called **USA State Plane Zones NAD83** and open it in Map Viewer.

4. Zoom to a familiar city or county and click a zone to determine its zone abbreviation and number.

5. Close your web browser.

Add a new layer to set a map's coordinate system

ArcGIS Pro offers several options for setting a map's coordinate system. One option is to add a layer with a coordinate system to a blank map; this layer's coordinate system will then be applied to the map and determine the map's coordinate system. Another option is to set a default coordinate system for all new maps in a project. Regardless of any coordinate system that has been applied, you can always change it on the map properties menu. In this section, you will use the first layer added to a map to set the coordinate system—in this case, State Plane Pennsylvania South.

1. Open **Tutorial5-3.aprx** from **Chapter5\Tutorials** and save it as **Tutorial 5-3YourName.aprx**.

   The project opens with the map **Allegheny County State Plane** and a basemap.

2. In the **Contents** pane, right-click **Allegheny County State Plane** > **Properties** > **Coordinate Systems**.

   Under **Current XY**, confirm that the default coordinate system, **WGS 1984 Web Mercator (Auxiliary Sphere)**, is being used. Next, you will confirm that the setting is on that changes the map to the coordinate system of the first layer added to the map.

3. On the **Project** tab, click **Options** > **Map and Scene**, expand **Spatial Reference**, and confirm that **Use spatial reference of first operational layer** is selected. Click **OK**.

This is where you can set the default coordinate system, using the **Choose spatial reference** option regardless of what layers you add to it. Next, you will add a layer with a state plane coordinate system and then a layer using a geographic coordinate system.

4. Click the back button, and on the **Map** tab, in the **Layer** group, click **Add Data**, browse to **Chapter5.gdb**, and add **Municipalities**.

   This layer for the municipalities of Allegheny County, Pennsylvania, uses **State Plane Pennsylvania South 3702** coordinates and the unit **US Feet**. Next, you will confirm that the map coordinate system changed when you added the layer.

5. In the **Contents** pane, right-click **Allegheny County State Plane** > **Properties** > **Coordinate Systems**, confirm under **Current XY** that **NAD 1983 State Plane Pennsylvania South FIPS 3702 (US Feet)** is displayed, and click **OK**.

   The coordinate system from the layer has been adopted in the map.

Add a layer that uses geographic coordinates

1. Add **Tracts** from **Chapter5.gdb**.

2. In the **Contents** pane, right-click **Tracts** > **Properties** > **Source** and expand **Spatial Reference**.

   The coordinate system for this layer is **GCS NAD 1983**. This layer uses geographic coordinates but is now projected on the fly to the map's coordinate system.

3. Move **Municipalities** above **Tracts** in the **Contents** pane.

4. Change the symbology of **Municipalities** to **No Color**, black outline, and width to **1.5**. Change **Tracts** to white, 30% gray outline, and width to **1** pt, and save your project.

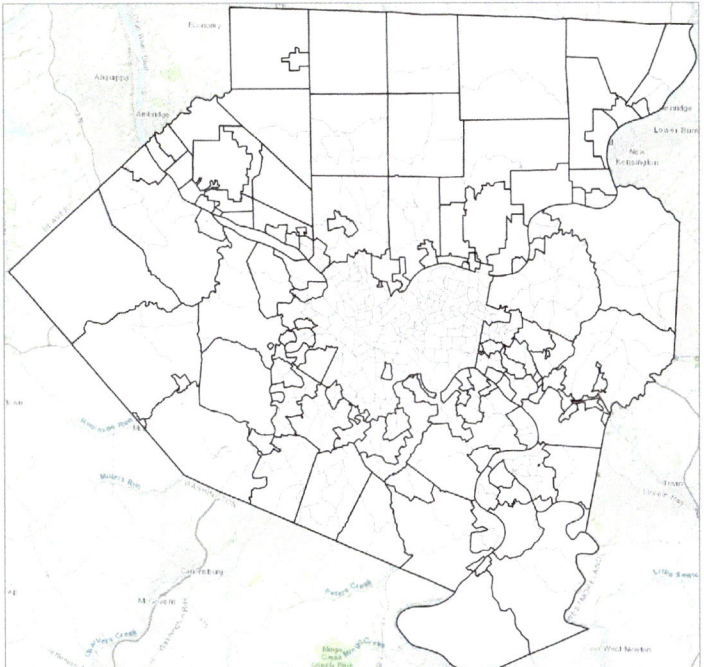

Change a map's coordinate system

The US military developed the Universal Transverse Mercator (UTM) grid coordinate system in the late 1940s. It covers the world with 60 longitudinal zones defined by meridians (longitude lines) that are six degrees wide. These zones, like state plane, are good for areas about the size of a state (or smaller).

1. Close the **Allegheny County State Plane** map and open the **California UTM** map.

   Both the map and the layer use geographic coordinates.

2. In the **Contents** pane, right-click **California UTM > Properties > Coordinate Systems**.

3. Expand **Projected Coordinate System > UTM > North America > NAD 1983**, click **NAD 1983 UTM Zone 11N**, and click **OK**.

   The map coordinates are in decimal degrees. Next, you will change these coordinates to meters.

4. In the **Contents** pane, right-click **California UTM > Properties > General**. Change **Display units** to **Meters** and click **OK**.

5. Zoom to the **California** bookmark and point to the center of the state.

The x,y coordinates at the bottom of the display are now in meters.

> **YOUR TURN**
>
> On the **Project** tab, click **Options** > **Map and Scene** > **Spatial Reference**. Click **Choose spatial reference**, browse to **Projected Coordinate System** > **World** > **WGS 1984 Web Mercator (auxiliary sphere)**, and click **OK**. Insert a new map, add **Counties**, and then view the coordinate system of the map. It's Web Mercator, the preferred format for ArcGIS Online. Save your project.

Tutorial 5-4: Working with vector data formats

This tutorial reviews file formats commonly found for vector spatial data, in addition to file geodatabases covered in chapter 4. Included are Esri shapefiles, x,y data, and Google Keyhole Markup Language (KML) files.

Examine a shapefile

Many spatial data suppliers use the shapefile data format, an Esri legacy format, for vector data because it's so simple. A shapefile consists of at least three files with the following extensions: .shp, .dbf, and .shx. Each file uses the shapefile's name but with a different extension (for example, Cities.shp, Cities.dbf, and Cities.shx). The .shp file stores the geometry of the features, the .dbf file stores the attribute table, and the .shx file stores an index of the spatial geometry. Next, you will examine a shapefile in more detail.

1. In File Explorer, browse to **Chapter5\Data**.

    CouncilDistricts is a shapefile for New York City Council Districts provided by the city's Planning Department. Shapefiles have three to seven associated files, including the three mentioned previously (.shp, .shx, and .dbf) and a PRJ file that contains the coordinate system.

2. Close File Explorer.

Import a shapefile into a file geodatabase and add it to a map

You will use a conversion tool to convert a shapefile into a feature class and store it in a file geodatabase.

1. Open **Tutorial5-4.aprx** from **Chapter5\Tutorials** and save it as **Tutorial 5-4YourName.aprx**.

   The project opens with a map zoomed to New York City with a projected coordinate system of **NAD 1983 State Plane New York Long Isl FIPS 3104 (US Feet)**.

2. Search for and open the **Export Features** tool. Apply the following settings and run the tool:
   - **Input Features**: CouncilDistricts.shp (located in Chapter5\Data)
   - **Output Feature Class: CouncilDistricts** (save in Chapter5.gdb)

   You now have a **CouncilDistricts** feature class in **Chapter5.gdb** that was automatically added to the **Contents** pane. The council districts accurately overlay the basemap.

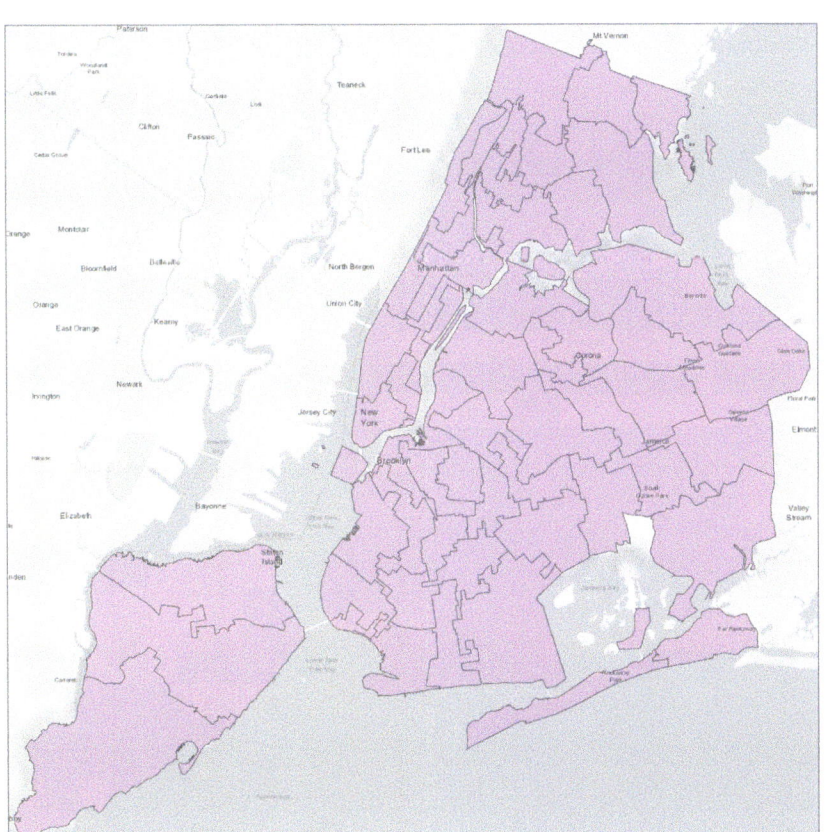

Add x,y data

Global Positioning System (GPS) units and many databases provide spatial coordinates as x,y coordinates. You can add these to a map. In this section, you will use x- and y-values from a table to create a point feature class of public libraries in New York City.

1. On the **Map** tab, click **Add Data**, browse to **Chapter5\Data**, and add **Libraries.dbf**.

2. In the **Contents** pane, right-click **Libraries** and click **Open**. Scroll to the right to see columns **XCOORD** and **YCOORD**.

   Locations are often formatted by using X to indicate longitude and Y to indicate latitude and are obtained from GPS units using the WGS84 geographic coordinate system. The coordinates in this table use the state plane coordinate system for New York City.

3. Close the table.

4. In the **Contents** pane, right-click **Libraries** and click **Create Points From Table > XY Table To Point**. Apply the following parameters:
   - **Output Feature Class**: **Libraries** (save in Chapter5.gdb)
   - **X field**: XCOORD
   - **Y field**: YCOORD
   - **Coordinate System**: Current Map [Vector Data Formats]

   The spatial reference **NAD_1983_StatePlane_New_York_Long_Island_FIPS_3104_Feet** is applied based on the map's setting.

5. Click **OK**.

The libraries appear as points on the map.

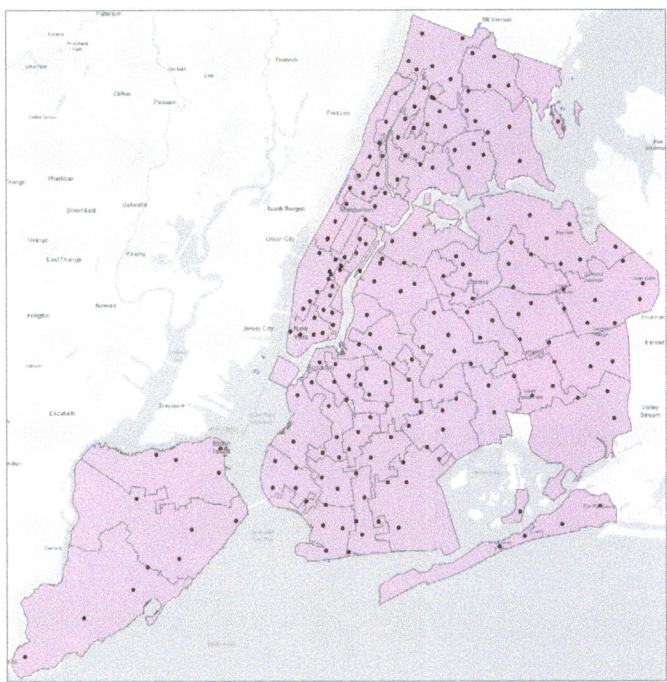

6. Remove the **Libraries** table from the **Contents** pane and save your project.

Convert a KML file to a feature class

KML is the file format used to display geographic data in many mapping applications. KML has become an international standard maintained by the Open Geospatial Consortium. In this section, you will convert the New York State Unified School Districts from a KML file to a feature class, stored in its own geodatabase.

1. Search for and open the **KML To Layer** tool. Apply the following parameters:
   - **Input KML File**: cb_2021_36_unsd_500k.kml (located in Chapter5\Data)
   - **Output Location**: Chapter5/Data
   - **Output Data Name**: **NYSchoolDistricts**

2. Run the tool.

   School district boundaries for the entire state are added. A file geodatabase with the specified name and a layer are created in the **Data** folder.

3. Remove the fill color for the **NYSchoolDistricts** polygon and choose a dark color for **Libraries**. Save and close your project.

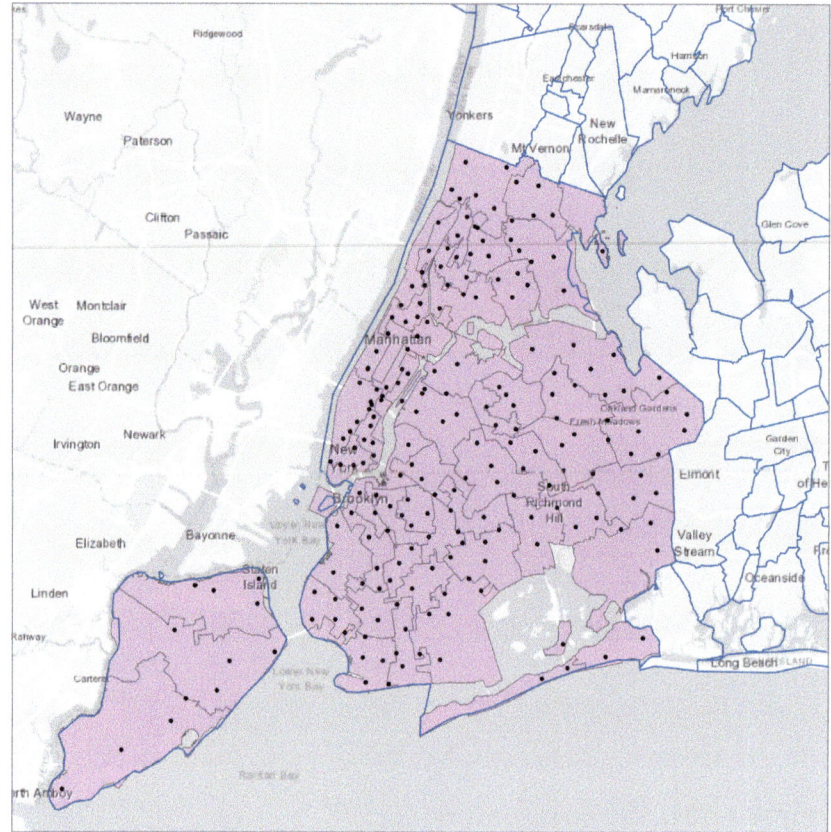

Tutorial 5-5: Working with US Census map layers and data tables

A large and ever-expanding collection of spatial data—including vector layers, raster layers, and data tables with geocodes—is available for free download from the internet or for direct use as services. The advantage of downloading spatial data is that you have more flexibility in modifying it. In this tutorial, you will learn about and use data from the US Census Bureau, a major spatial data supplier.

From the Census Bureau's website, you can download polygons with fields containing unique identifiers for geographic areas, identifying census tracts, for example, that don't include fields for census attributes such as population, age, sex, race, income, and so on. Because there are thousands of census variables, it is impractical to include all of them in attribute tables. The layers would be excessively large and cumbersome files. Instead, the Census Bureau provides websites in which you can build custom collections of variables for your needs. The Census Bureau continually updates these websites, so you may find that some popular collections of data are already available and that the website's menus vary.

In this tutorial, you will download a Topologically Integrated Geographic Encoding and Referencing (TIGER) shapefile for Minnesota and edit it to show Hennepin County, where the city of Minneapolis is located. You will then select census demographic data and download it, perform data preparation steps, and join the data to the corresponding polygon boundary layer.

Download census TIGER files

US Census boundaries are redrawn periodically, so it's important to match the year the boundary files were published with the year the demographic data was collected. In this case, you will research 2020 data.

1. In a web browser, go to **census.gov/cgi-bin/geo/shapefiles**.

   There are alternative methods for downloading TIGER files as file geodatabases. This is a direct link to individual shapefiles.

2. Choose **2020** for **Select year** and **Census Tracts** for **Select a layer type**. Click **Submit**.

3. Under **2020 Census Tracts**, for **Select a State**, click **Minnesota**, and click **Download**.

   This selection downloads a shapefile of all census tracts for the entire state. You will later limit your area of interest to Hennepin County using ArcGIS Pro.

4. When the download completes, extract the file to **Chapter5\Tutorials\Downloads\Census**.

   > **YOUR TURN**
   >
   > In your browser, go back to the first web page and download the **2020 County Subdivisions** for **Minnesota**. Return to the first web page and choose **2020** for the year and **Water** for the layer type. Under **Area Hydrography**, choose **Minnesota** and then **Hennepin County**, and download it. Extract both files to the **Chapter5\Tutorials\Downloads\Census** folder.

Download census tabular data

The Census Bureau provides detailed data on education, income, transportation, and other subjects. In this section, you will download commuting-to-work data at the tract level for Hennepin County, Minnesota. The city of Minneapolis is ranked high in the nation for the quality of its transportation infrastructure, including bike lanes, according to several groups and publications. The census is a good starting point to analyze means of transportation to work by male, female, and overall percentages of those who bicycle to work. You will first select the geography (all tracts in Hennepin County) and the topic (commuting characteristics by sex).

1. In a web browser, go to **data.census.gov** and click **Advanced Search**.

2. Click **Geographies** > **Census Tract** > **Minnesota** > **Hennepin County, Minnesota**, and click **All Census Tracts within Hennepin County, Minnesota**.

   This will add all tracts in this county to a selection.

3. Click **Search for a filter or table**. In the search box, type **Commuting**.

4. Under **Topic**, check the box for **Commuting**.

5. At the top of the page, click the **Table** tab.

   A list of related tables appears.

6. Click the first table, **Commuting Characteristics by Sex** (table **SO801**).

7. Click **DOWNLOAD TABLE** and modify the selection so that only the **2020** box is selected.

8. Click **DOWNLOAD .ZIP**.

9. When the download completes, extract the download to **Chapter5\Tutorials\Downloads\Census**.

Process tabular data in Microsoft Excel

The census data that you downloaded in the previous steps needs some cleaning up using Microsoft Excel before you use the data in a GIS. You will delete all columns except those needed for joining and the estimate of percentages of males and females who bicycle to work.

1. Browse to **Chapter5\Tutorials\Downloads\Census** and open **ACSST5Y2020.S0801-Data.csv**.

   The file opens in Microsoft Excel. View the extent of the columns by scrolling right.

2. Delete all fields except the following:
   - **Column A**: GEO_ID
   - **Column EG**: Estimate!!Male!!Workers 16 years and over!!MEANS OF TRANSPORTATION TO WORK!!Bicycle
   - **Column IQ**: Estimate!!Female!!Workers 16 years and over!!MEANS OF TRANSPORTATION TO WORK!!Bicycle

   *Tip: As you delete columns, the column label associated with an attribute will change. To prevent confusion, find **Column EG** and **Column IQ** and highlight the column heading. Then, starting from the right, while pressing **Shift**, select the columns to be deleted and delete the columns in batches. The highlighted columns will help you figure out where to stop your selection.*

3. Delete the first row.

4. Rename the columns: **GEOID**, **MALE_BIKE**, and **FEMALE_BIKE**.

5. Select column **GEOID**.

6. On the **Home** tab, in the **Editing** group, click **Find & Select** and click **Replace**.

7. In the **Find and Replace** dialog box, for the **Find what** field, type **1400000US**, leave the **Replace with** field blank, click **Replace All**, and click **Close**.

   This converts the values in the census tract **GEOID** field to a number that includes just the tract numbers. The field type is not yet numeric, so you will need to change its format.

8. Select the columns **MALE_BIKE** and **FEMALE_BIKE**, right-click, and click **Format Cells**.

9. In the **Format Cells** dialog box, under **Category**, click **Number**, change **Decimal places** to **1**, and click **OK**.

10. Highlight the column **GEOID**, right-click, and click Format Cells. Click **Number** and change **Decimal places** to **0**. Click **OK**.

	A	B	C
1	GEOID	MALE_BIKE	FEMALE_BIKE
2	27053000101	0.0	0.0
3	27053000102	0.9	0.0
4	27053000300	1.7	0.0
5	27053000601	4.6	1.7
6	27053000603	0.3	0.0

11. Save the file as **BikeWorkData.csv**.

Add and clean data in ArcGIS Pro

1. Open **Tutorial5-5.aprx** from **Chapter5\Tutorials** and save it as **Tutorial 5-5YourName.aprx**.

2. Using the **Export Features** tool, import the three census shapefiles in your **Chapter5\Tutorials\Downloads\Census** folder to **Chapter5.gdb** and set these parameters, as shown in the table:

Input Features	Output Feature Class
tl_2020_27_cousub	HennepinCountySubdivisions
tl_2020_27_tract	MinnesotaTracts
tl_2020_27053_areawater	HennepinWater

3. Use the **Table to Geodatabase** tool to convert **BikeWorkData.csv** into the output geodatabase **Chapter5.gdb**.

4. In the **Catalog** pane, expand **Databases** and open **Chapter5.gdb**. Right-click **BikeWorkData** and click **Add To Current Map**.

5. Open the table for **BikeWorkData** and sort by ascending order for the **GEOID** field.

   GEOID is a numeric field.

6. Close the **BikeWorkData table** and open the attribute table for **MinnesotaTracts**.

   The candidate field for joins is **GEOID** and is a text field.

7. Open the **Fields** view and create a numeric field (with **Double** data type) named **GEOIDNUM**.

8. Save the changes and close the **Fields** view.

9. Use the **Calculate Field** tool to make **GEOIDNUM** equal to **GEOID**.

   The join field in both tables of a join must be the same type. This step calculates values in the **GEOIDNUM** field equal to what is in the **GEOID** field, but these values will now be stored as a numeric data type. GEOIDNUM can now be joined to the numeric data type field used in the **BikeWorkData** table.

10. Close the **MinnesotaTracts** table.

Join data and create a choropleth map

The data for bicycling to work is only for Hennepin County. When the state census tracts are joined to it, you will adjust a setting to retain only the fields that can be joined. This will result in only Hennepin County bike usage data.

1. Use the **Add Join** tool to join the **BikeWorkData** table to **MinnesotaTracts** using fields **GEOIDNUM** for the **Input Join** field and **GEOID** for the **Join Table** field. Uncheck the box for **Keep all input records**.

   This will keep only Hennepin County tracts.

2. Export the features as **HennepinTractsBikeWork** to **Chapter5.gdb**. Copy and paste the layer so there are two copies in the **Contents** pane.

3. Rename one layer **Male bicyclists** and the other layer **Female bicyclists**.

   The data fields are already percentages.

4. Remove the original **MinnesotaTracts** layer and **BikeWorkData** table.

5. Zoom to the **Female bicyclists** layer and symbolize it using **Graduated Colors**.

6. For **Field**, click the **Set an expression** button to open the **Expression Builder**. Delete the current expression, and in the **Fields** list, double-click **FEMALE_BIKE**. Click **OK**.

7. Symbolize using the color scheme **Oranges (5 classes)**, and manually set the class upper values to **1.5**, **3**, **6**, **12**, and **13.3**. On the **Advanced symbology options** tab, expand **Format labels**. Change the **Category** to **Percentage** and **Decimal places** to **1**.

8. Similarly, symbolize the **Male bicyclists** layer using the MALE_BIKE field and the color scheme **Purples (5 classes)**, and manually set the class upper values to **1.5**, **3**, **6**, **12**%, and **18.3**%. Apply the same legend label formatting settings as in step 6.

9. Move **HennepinCountySubdivisions** to the top of the **Contents** pane, change its fill to **No color**, and set a black **Outline width** of **1.5**.

10. Move **HennepinWater** below **HennepinCountySubdivisions** and change its symbology to **Water (area)**.

11. Turn the layers on and off to see the difference in percentage of males versus females who bicycle to work, and then save your project.

    Do males or females bicycle to work in different parts of Minneapolis at a higher percentage?

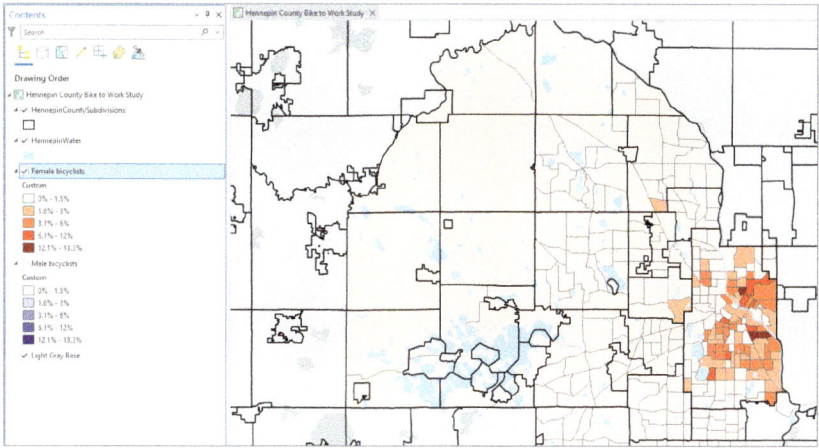

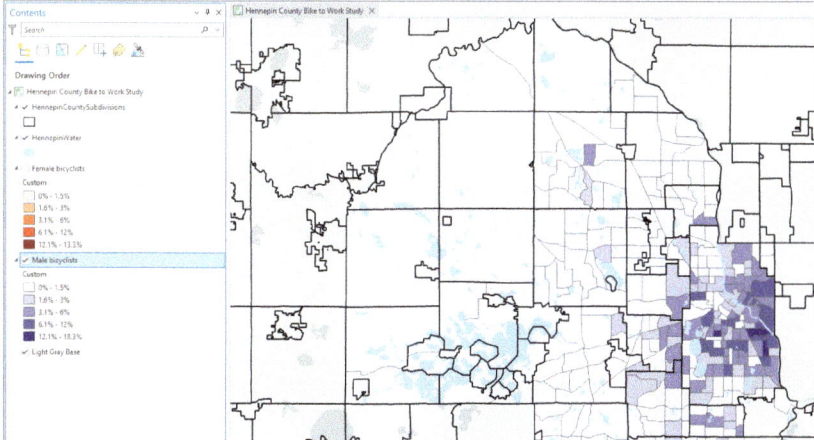

Tutorial 5-6: Downloading geospatial data

There are many government websites, such as the USGS's National Map Viewer or Data.gov, where you can download geospatial data for use in a GIS. Other agencies that create data include the US Department of Agriculture (USDA), Department of Commerce (DOC), National Oceanic and Atmospheric Administration (NOAA), US Census Bureau, Department of the Interior (DOI), Environmental Protection Agency (EPA), and National Aeronautics and Space Administration (NASA). Many local agencies provide their data on open data websites such as NYC Open Data or the Western Pennsylvania Regional Data Center.

You can access and quickly use data from many of these organizations and other content from ArcGIS Living Atlas. Its geospatial data includes maps and data on thousands of topics in the form of imagery, basemaps, and features that can be added directly or saved locally. Examples include historic maps, demographics for the United States and many other countries, landscapes, oceans, Earth observations, urban systems, transportation, boundaries, and places. You can use and contribute data, maps, or apps to ArcGIS Living Atlas. Data is frequently added, so see livingatlas.arcgis.com for current offerings.

Add a land-use layer from ArcGIS Living Atlas

In this section, you will search for and add a land-use raster layer from ArcGIS Living Atlas and extract it for only Hennepin County, Minnesota. Whereas vector maps are discrete—consisting of points, lines, and polygons connecting coordinates—rasters, such as the one used in this tutorial, represent continuous phenomena. Rasters use many of the same file formats as images on computers, including JPEG (.jpg) and TIFF (.tif) formats. All rasters are rectangular, consisting of rows and columns of cells known as pixels. Each pixel has associated x,y coordinates and a z-value attribute, such as altitude for elevation or some other property. In ArcGIS Pro, you can add data from ArcGIS Living Atlas using the **Catalog pane** > **Portal** > **Living Atlas** or using the **Add Data** button.

1. Open **Tutorial5-6.aprx** from **Chapter5\Tutorials** and save it as **Tutorial 5-6YourName.aprx**.

2. On the **Map** tab, click **Add Data**. Click **Portal** > **Living Atlas**.

3. In the **Search Living Atlas** box (top-right corner), type **NLCD** and press **Enter**.

    This search will retrieve items from the National Land Cover Database.

4. Click **USA NLCD Land Cover** and click **OK**.

    Tip: If you can't see the title of the layer, click on it and the name will appear in the **Name** box.

    The raster is added to the map of Hennepin County, Minnesota, and includes a legend showing land use.

Extract raster features for Hennepin County

A key aspect of rasters is that they are large files. So, although you may store some important basemap raster files on your computer, these maps are perhaps best obtained as map services available for display on your computer but stored elsewhere such as ArcGIS Living Atlas. If you want to extract a subset of the land use for one county and store the raster file on your local computer, you must use the **Extract by Mask** tool.

1. Search for and open the **Extract by Mask** tool. If a **Tool Not Licensed** error is generated, review the software requirements and licensing section at the beginning of the book for authorizing this tool.

2. On the **Environments** tab, under **Processing Extent**, for **Extent**, click the **Extent from a Layer** button and select **HennepinCounty** from the list.

3. On the **Parameters** tab, for **Input Raster**, choose **USA NLCD Land Cover**, and for **Input raster or feature mask data**, choose **HennepinCounty**. Save the output raster as **HennepinCountyLandUse** in **Chapter5.gdb**.

4. Run the tool.

5. Remove the **USA NLCD Land Cover** layer, zoom to the **HennepinCountyLandUse** layer, and save your project.

   Notice the differences in Hennepin County with developed areas on the eastern side of the county and forest, pasture or hay, and cultivated crops in the western part of the county.

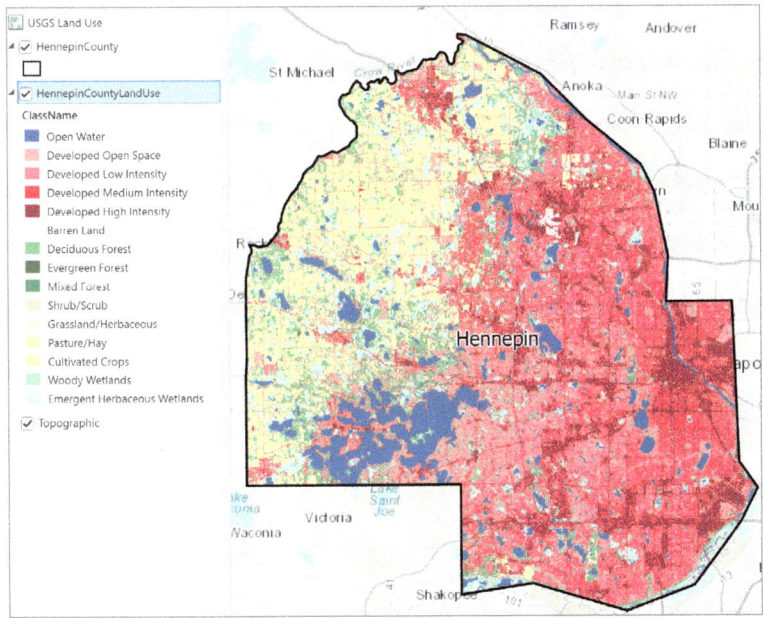

YOUR TURN

Search for topics (for example, health or housing) or geographic areas. Add and explore the contents of various ArcGIS Living Atlas layers.

Download elevation contours from a government organization

Although data from organizations such as the USGS is available in ArcGIS Living Atlas, you can download features directly from government websites. Walkability and bicycling studies often require knowledge of the topography of a city, and elevation contours are needed. You can get these contours from the USGS.

1. In a web browser, go to **apps.nationalmap.gov/downloader**.

2. In the map's search box, search for **Minneapolis, MN, USA** or zoom to the area.

3. On the **Datasets** tab, under **Data**, check the box for **Elevation Products (3DEP)** and leave only the box for **Contours (1:24,000-scale)** checked.

4. Under **File Formats**, confirm that **Shapefile** is selected. Scroll up and click **Search Products**.

5. Find the choice for **Saint Paul W, Minnesota**. Click the **Add to Cart** button.

6. On the **Cart** tab, under **Download**, click the link to download the file.

   Wait while the file downloads.

7. Locate the downloaded file and extract its contents to **Chapter5\Tutorials\Downloads\USGS**.

8. In ArcGIS Pro, open the **USGS Contours** map and zoom to the **Minneapolis** layer.

9. Click **Add Data** and browse to **Chapter5\Tutorials\Downloads\USGS**. Open the subfolder that was created and add the **Elev_Contour.shp** file to the map.

10. In the **Contents** pane, drag **Elev_Contour** below Minneapolis.

    The map displays contours of the city. To the northeast, you can see areas with few contours, indicating that they are relatively flat. Later in the book, you will learn how to clip these features to the city's boundary.

11. Symbolize the contours as ground features and save your project.

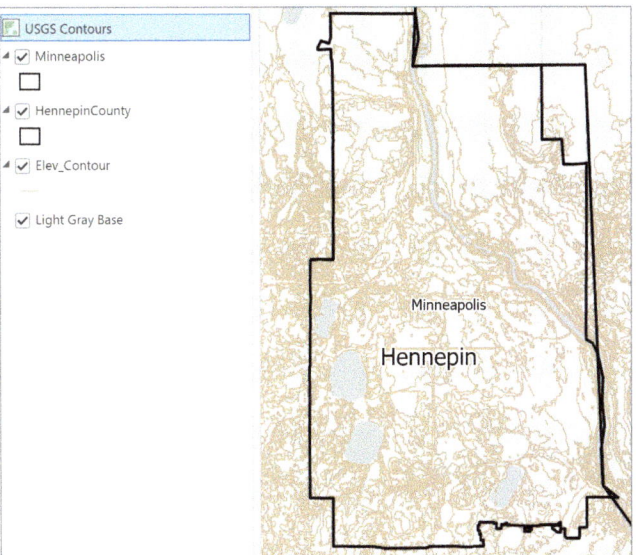

Download local data from a public agency hub

Many local agencies and GIS departments provide GIS data for little or no cost through open data portals or hubs. You will download bike-related data from Hennepin County's open data hub. More information about the county's biking transportation is available on the hub's website.

1. In a web browser, go to **gis-hennepin.hub.arcgis.com**, click the **Open Data** tab, and click **Transportation**.

   The resulting list shows transportation layers available for use in GIS. Layers such as **Hennepin County Bike and Pedestrian System** would be useful for a bike-to-work study.

2. Click **Hennepin County Bike and Pedestrian System**.

   A map appears with the location of bike and pedestrian routes. This data can be downloaded for direct use in GIS.

3. Click **Download**.

   Data type options include **CSV**, **KML**, **Shapefile**, or **GeoJSON**.

4. Under **Shapefile**, click **Download**. Extract it to **Chapter5\Tutorials \Downloads\OpenDataHub**.

5. In ArcGIS Pro, open the **Hennepin Bike and Ped System** map.

6. Add the **Hennepin_County_Bike_and_Pedestrian_System** and symbolize using **Unique Values** and the **Facility** field.

7. Change the projected coordinate system of the map to **NAD 1983 (2011) (US Feet) State Plane Minnesota South FIPS 2203 (US Feet)**.

The line features can be used with the US Census bike-to-work data from tutorial 5-5 to analyze bicycle ridership and routes.

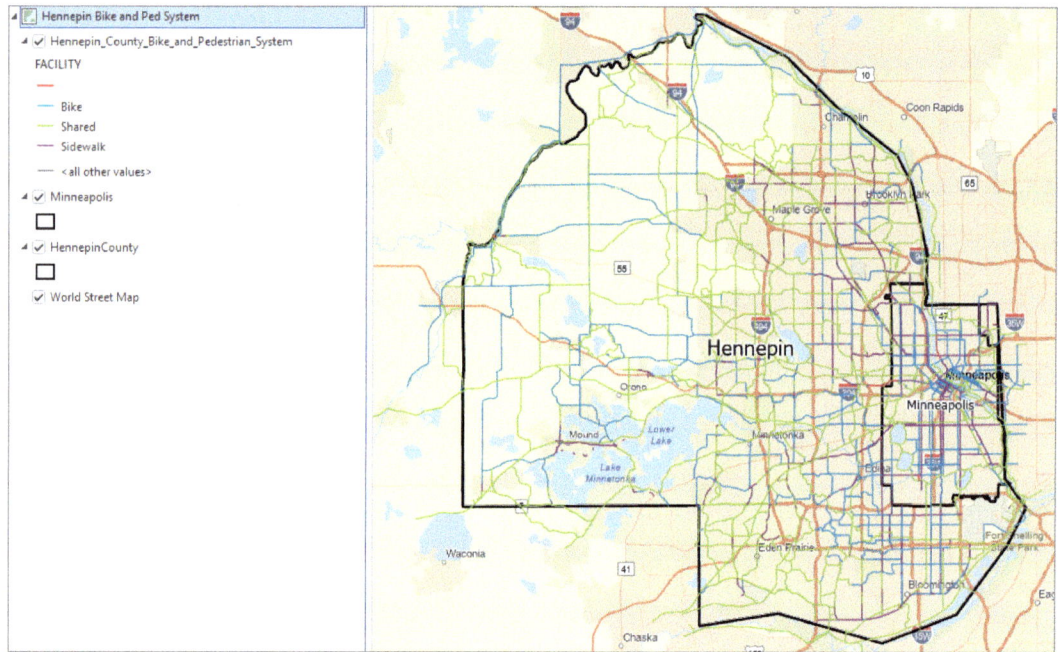

YOUR TURN

Search for and explore public datasets of interest from Hennepin County or other local agencies' open data hubs.

Assignments

This chapter has assignments to complete that you can download with data from ArcGIS Online at links.esri.com/GISTforPro3.4Assignments.

CHAPTER 6

Geoprocessing

LEARNING GOALS

- Dissolve block group polygons to create neighborhoods and fire battalions and divisions.
- Extract a neighborhood using attributes to form a study area.
- Extract features from other layers using the study area.
- Merge water features to create a single water map.
- Append separate fire and police station layers to one layer.
- Intersect streets and fire companies to assign street segments to fire companies.
- Use Union on neighborhood and land-use boundaries to create detailed polygons of neighborhood land-use characteristics.
- Apportion data between two polygon layers whose boundaries do not align.

Introduction

Geoprocessing is a framework and set of tools for processing geographic data. Generally, you must use geoprocessing tools to build study areas in a GIS and perform tasks. In this chapter, you will learn how to extract a subset of spatial features from a map using attribute or spatial queries; aggregate polygons into larger polygons; append layers to form a single layer; and use Intersect, Union, and Tabulate Intersection tools to combine features and attribute tables for geoprocessing. In this chapter, for example, you'll process and prepare layers for emergency management officials in New York City's Manhattan borough and one of its neighborhoods, the Upper West Side.

Tutorial 6-1: Dissolving features to create neighborhoods and fire divisions and battalions

Suppose emergency response planners want to know the number of housing structures or units by neighborhood for planning purposes. New York City neighborhoods are composed of block groups that include housing data. In this tutorial, you'll create neighborhood divisions within Manhattan by dissolving block groups within the neighborhood. For each neighborhood, dissolving retains the neighborhood's outer boundary lines but removes interior lines from the block groups. The **Pairwise Dissolve** tool can aggregate block group attributes to the neighborhood level, using statistics such as sum, mean, and count. In this case, you'll sum the number of structures in housing units by neighborhood.

Examine the Dissolve field and other attributes

The **Pairwise Dissolve** tool needs data—a **Dissolve** field, for combining block groups into a neighborhood. In this case, **Name** is the **Dissolve** field, and that field contains the neighborhood name to which each block group belongs.

1. Open **Tutorial6-1.aprx** from the **Chapter6\Tutorials** folder and save the project as **Tutorial6-1YourName.aprx**.

   The map contains Manhattan block groups with housing units and structures.

2. Open the **ManhattanBlockGroups** attribute table and sort the **Name** field in ascending order.

   As you scroll, you can see the block group records that make up each neighborhood. The figure shows some of the block groups for the **BatteryParkCity-LowerManhattan** neighborhood and their associated **GEOID**.

	OBJECTID *	Shape *	STATEFP	COUNTYFP	GEOID	Name
1	18	Polygon	36	061	360610013002	BatteryParkCity-LowerManhattan
2	21	Polygon	36	061	360610015011	BatteryParkCity-LowerManhattan
3	22	Polygon	36	061	360610015021	BatteryParkCity-LowerManhattan
4	107	Polygon	36	061	360610007001	BatteryParkCity-LowerManhattan
5	111	Polygon	36	061	360610009001	BatteryParkCity-LowerManhattan

3. Scroll to the right and examine the remaining attributes.

Values are an estimate of the number of housing units in structures of different sizes, along with the total number of housing units in each block group. For example, the field **TOT1_Detached** is a one-family house detached from any other house. **TOT20_49** includes the total number of housing units for each block group in structures containing 20 to 49 units (that is, larger apartment or condominium buildings). **TOT50** is the total number of housing units in structures containing 50 or more units (mainly high-rise apartment and condo buildings). When dissolving block groups, you'll sum these housing unit values by neighborhood, which will indicate housing density.

4. Close the table.

Dissolve block groups to create neighborhoods

1. Search for and open the **Pairwise Dissolve** tool and apply these settings:
   - **Input Features**: ManhattanBlockGroups
   - **Output Feature Class**: **ManhattanNeighborhoods**
   - **Dissolve Field**: Name
   - **Field**: all the fields between TOT1_ATTACHED to TOT50 (eight total)
   - **Statistic Type:** Sum

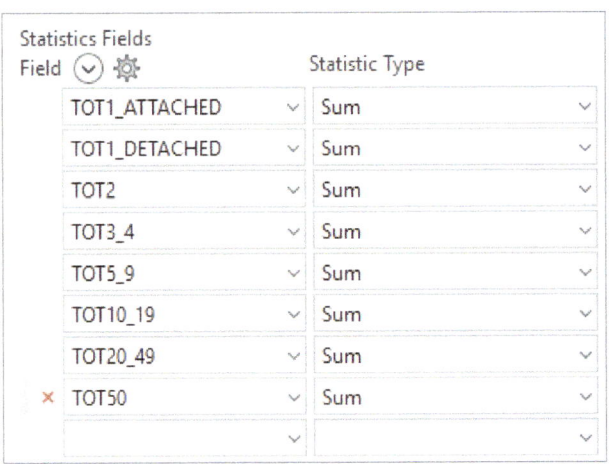

2. Run the tool.

3. In the **Contents** pane, open the **ManhattanNeighborhoods** attribute table and sort the **Sum_TOT50** field in descending order.

For example, on the Upper West Side, there are 41,240 housing units in structures having 50 or more units.

OBJECTID	Shape	Name	SUM...	SUM...	SUM...	SUM...	SUM...	SUM...	SUM...	SUM_TOT50
1	23	Polygon	Upper West Side	543	500	481	1499	6093	8810	15806	41240
2	12	Polygon	Lenox Hill - Roosevelt Island	246	222	204	483	2025	5080	8785	34976
3	13	Polygon	Lincoln Square	211	207	275	244	1609	2554	4480	31340
4	22	Polygon	Turtle Bay - East Midtown	70	268	176	268	750	1664	4382	29618
5	28	Polygon	Yorkville	222	391	85	570	1543	9061	9510	27618

4. Close the tables and map and open the **Fire Companies Battalions and Divisions** map.

Other useful layers for city officials and emergency response teams would show population per square mile for fire battalions and identify fire divisions. The New York City Fire Department, like most fire departments around the country, is organized in a quasi-military fashion with companies, battalions, and divisions. In the **Fire Companies Battalions and Divisions** map, you have fire companies but need fire battalions and divisions.

> **YOUR TURN**
>
> Use the **Pairwise Dissolve** tool on **ManhattanFireCompanies** and the dissolve field **FireBN** to dissolve polygons and create a feature class named **ManhattanFireBattalions**. Sum the attributes **TOT_POP** and **SQ_MI** to find the population and square miles, respectively, for each battalion. Perform another **Pairwise Dissolve** operation on **ManhattanFireCompanies**, with the same attributes to create a feature class named **ManhattanFireDivisions** using the dissolve field **FireDiv**.

5. Remove **ManhattanFireCompanies**.

6. Symbolize **ManhattanFireDivisions** by changing **Color** to **No color**, changing **Outline color** to red, and increasing **Outline width** to **2**.

7. Symbolize **ManhattanFireBattalions** with **Graduated Colors**. For **Field**, use **Sum_TOT_POP**. For **Normalization**, use **Sum_SQ_MI**. Set the **Method** to **Natural Breaks (Jenks)**. Change the **Color scheme** to **Blues (5 classes)**.

8. Label the fire battalions using the field **FireBN** to show the battalion number.

Is this one of the most densely populated fire battalion areas?

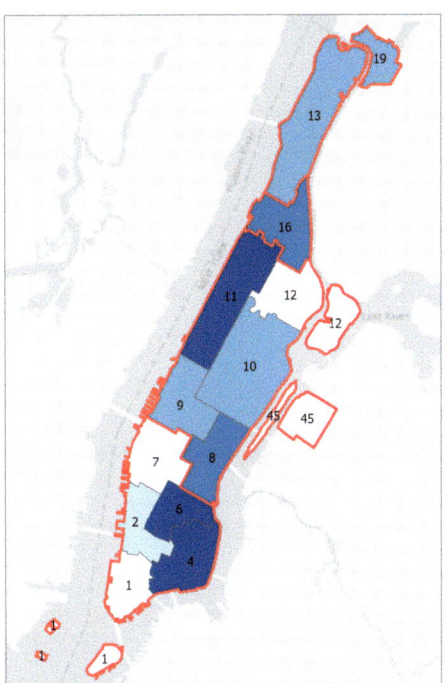

Tutorial 6-2: Extracting and clipping features for a study area

This tutorial is a workflow for creating a study region from layers that have more features than needed. Suppose you want to study housing units and streets for just one neighborhood. First, you will create the study area neighborhood by selecting the neighborhood using the attribute table and creating a single polygon for that study area. Second, you will use the new polygon and select by location to create features of block groups in the study area only. Third, you will use **Clip**, a geoprocessing tool, to clip streets to the study area.

Use Select By Attributes to create a study area

In this section, you will create a study area that includes a polygon feature of the Upper West Side neighborhood and block groups and streets for that neighborhood only. Such a study area is important when you work in geographic areas, such as New York City, that have many streets and block groups.

1. Open **Tutorial6-2.aprx** from the Chapter6\Tutorials folder and save the project as **Tutorial6-2YourName.aprx**.

The map contains New York City neighborhoods, block groups, and Manhattan streets.

2. Use the **Upper West Side** bookmark.

3. On the **Map** tab, in the **Selection** group, click **Select By Attributes** and apply these settings:
   - **Input Rows**: New York City Neighborhoods
   - **Expression**: Where Name is equal to Upper West Side

4. Click **OK**.

   The result shows the Upper West Side neighborhood selected. You will now export the feature to display and symbolize the Upper West Side neighborhood.

5. In the **Contents** pane, right-click **New York City Neighborhoods** and click **Data** > **Export Features**.

6. For **Output Feature Class**, type **UpperWestSide** and click **OK**.

7. Remove the **New York City Neighborhoods** layer and zoom to the **UpperWestSide** layer.

Use Select By Location to create study area block groups

In the next steps, you will use the **Select By Location** tool to select Manhattan block groups that belong to the Upper West Side neighborhood. After selecting block groups, you will create a feature class from the selected features. Because block groups are not contiguous with the neighborhood boundary, there are a few ways to create this selection. Here, you will select the block groups whose centers are in the Upper West Side neighborhood and manually select the remaining block groups that are partially inside the neighborhood.

1. On the **Map** tab, in the **Selection** group, click **Select By Location** and apply these settings:
   - **Input Features**: New York City Block Groups
   - **Relationship**: Have their center in
   - **Selecting Features**: UpperWestSide

2. Click **OK**.

   A few more block groups need to be manually selected.

3. On the **Map** tab, click the **Select** button. While pressing the **Shift** key, select the remaining block groups on the west (left) side of the neighborhood, outside the outline.

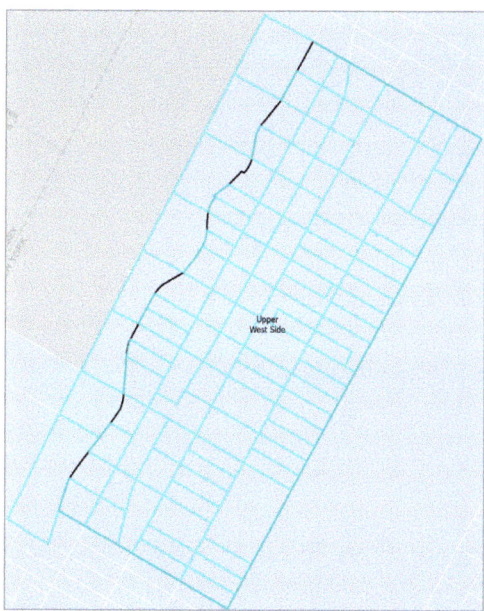

4. Using the **Export Features** tool, export the selected features as **UpperWestSideBlockGroups** to **Chapter6.gdb**.

   Tip: The selected features are within the **New York City Block Groups** layer.

5. Remove the original **New York City Block Groups** layer, move the **UpperWestSide** layer to the top of the **Contents** pane, and turn on **Manhattan Streets**.

> ### YOUR TURN
> Use the **Select By Location** tool to select **Manhattan Streets** that intersect the **UpperWestSide** layer. Notice that many of the streets dangle, or extend beyond the polygon boundary. These selected streets, each a full block long, are needed for geocoding address data to points, as explained in chapter 8. Save the selection as **UpperWestSideStreetsForGeocoding**. Remove the **Manhattan Streets** layer, turn off **UpperWestSideBlockGroups**, and move **UpperWestSide** to the top of the **Contents** pane.

Clip streets

Next, you will use the **Pairwise Clip** tool to cleanly create street segments using the Upper West Side polygon. Clipping Manhattan Streets with the Upper West Side boundary cuts off the dangling portions of the streets and creates a clean layer for display purposes.

1. Search for and open the **Pairwise Clip** tool. Apply these settings and run the tool:
    - **Input Features**: UpperWestSideStreetsForGeocoding
    - **Clip Features**: UpperWestSide
    - **Output Feature Class**: **UpperWestSideStreets**

2. Turn off **UpperWestSideStreetsForGeocoding** and **UpperWestSideBlockGroups**.

    The streets are cut cleanly to the Upper West Side neighborhood.

Tutorial 6-3: Merging water features

Sometimes, you must merge two or more layers into a new single layer. In this section, you will build one water feature class from several adjacent water feature classes. New York City is made up of five boroughs, each of which is also a county whose water features are downloaded separately from the US Census Bureau.

Merge features

You will use the **Merge** geoprocessing tool to create one water feature class from five separate feature classes.

1. Open **Tutorial6-3.aprx** from the **Chapter6\Tutorials** folder and save the project as **Tutorial6-3YourName.aprx**.

   The map, **NYC Water**, includes separate water features for each borough.

2. Search for and open the **Merge** tool. Apply these settings and run the tool:
   - **Input datasets**: BronxWater, BrooklynWater, ManhattanWater, QueensWater, and StatenIslandWater
   - **Output Dataset**: **NYCWater**

3. Turn off the original water layers so that only the new **NYCWater** layer is visible.

4. For **NYCWater**, change the **Color** to blue and zoom to the layer.

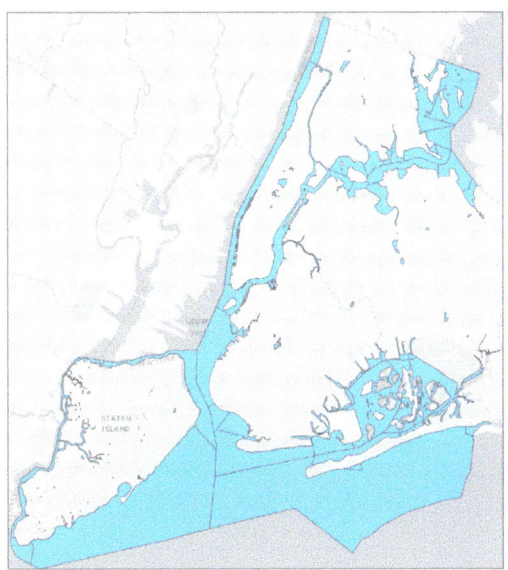

5. Open the attribute table for **NYCWater**.

All the polygons from the source layers have been merged into one feature class, resulting in a water feature class with 330 records. If you want to create a single polygon, you must do a dissolve.

> **YOUR TURN**
>
> Open the **NYC Waterfront Parks** map and merge the separate feature classes for each borough into one feature class named **NYCWaterfrontParks**, with a green color.

Tutorial 6-4: Appending firehouses and police stations to EMS facilities

City agencies may track facility locations in separate feature classes and tables, but suppose that an emergency medical services (EMS) official wants them in one feature class for better planning. You will use the **Append** tool to add features to an existing feature class, considering that both have the same attributes (or same schema). In this tutorial, you will add firehouse and police station points to a feature class named **EMS Facilities**.

Append features

1. Open **Tutorial6-4.aprx** from the **Chapter6\Tutorials** folder and save the project as **Tutorial6-4YourName.aprx**.

   The map contains EMS facilities.

   You will use the **Append** tool to append firehouses and police stations to already existing EMS points (**EMS Facilities**). Each layer has the same table (field) structure, also known as a schema. This allows you to choose the option for matching the input table's schema to the target table's schema.

2. Open the **EMS Facilities** attribute table.

   Records for 34 facilities are included.

3. Search for and open the **Append** tool. Apply these settings and run the tool:
   - **Input Datasets**: FireHouses and PoliceStations (located in Chapter6.gdb)
   - **Target Dataset**: EMSFacilities

In the **EMS Facilities** attribute table, scroll to see the added firehouses and police stations.

Tutorial 6-5: Intersecting features to determine streets in fire company zones

New York City fire companies include engine, ladder, rescue, and squad companies, each with different roles. For example, engine companies secure water from a fire hydrant and extinguish a fire. Ladder companies do search and rescue, forcible entry, and ventilation at the scene of a fire. Rescue and squad companies are highly trained and deal with incidents that are beyond the duties of standard engine or ladder companies. Squad companies are also trained in mitigating hazardous materials that threaten Manhattan.

For response planning, each fire company must know the total length of streets they cover by the type of company (engine or ladder). To summarize streets for each company, streets must have the name and type of their fire company. The **Pairwise Intersect** tool achieves this summary by creating a feature class combining all the features and attributes of two input (and overlaying) feature classes—fire companies and streets. The **Intersect** tool excludes any parts of two or more input layers that don't overlie each other. Because fire companies have the same number in different fire battalions, a field that includes the battalion number and fire company is used in the calculations.

After streets are intersected, you can sum them by fire company type to determine how many streets are served by engine, ladder, or squad companies in Manhattan.

Open tables to study attributes before intersecting

1. Open **Tutorial6-5.aprx** from the **Chapter6\Tutorials** folder and save the project as **Tutorial6-5YourName.aprx**.

   The map contains New York City fire companies (polygons) and Manhattan streets (lines). Fire companies are classified by company type. **ManhattanStreets** is turned off.

2. Open the **ManhattanStreets** attribute table and review the fields.

   Studying the attribute tables of each feature class familiarizes you with the attributes before you intersect features. You will find no data about fire companies in the **ManhattanStreets** feature class, but you will find information about street characteristics, such as the shape (polyline) and length of each street segment.

3. Open the **ManhattanFireCompanies** attribute table and sort **FireCoNum** (fire company number) in ascending order.

Examine the attributes and data of this table. Fields of interest are the shape (polygon) and **FireBN_Co_Type**, a field that combines fire battalion, company number, and fire company type fields. For example, ladder company (**L**), engine company (**E**), or fire squad (**Q**) are fire company types.

4. Close the attribute tables.

Intersect features

1. Search for and open the **Pairwise Intersect** tool. Apply these settings and run the tool:
   - **Input Features**: ManhattanStreets and ManhattanFireCompanies
   - **Output Feature Class**: **ManhattanFireCompanyStreets**
   - **Output Type**: Line

2. Turn off all layers except **ManhattanFireCompanyStreets** and the basemap, zoom to the **Central Park** bookmark, and zoom in a few times.

   The result is street centerlines that include data about the fire companies serving each street.

3. Click the **Explore** button, click any of the line features of **ManhattanFireCompanyStreets**, and scroll to the bottom of the pop-up window.

The result is street centerlines that include data about the fire companies serving each street.

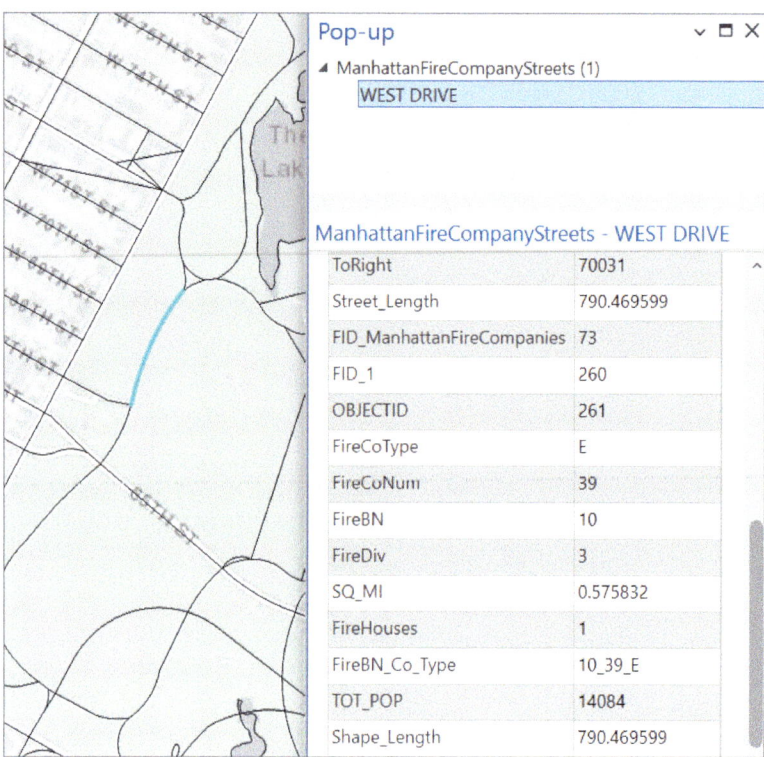

Summarize street length for fire companies

Now that each street segment has fire battalion, company, or type information, you can summarize this information so that emergency planners and fire officials know the total length of streets that each company and fire type services.

1. Open the **ManhattanFireCompanyStreets** attribute table, right-click **FireBN_Co_Type**, and click **Summarize**.

2. Apply these settings to the **Summary Statistics** tool and click **OK**:
   - **Input Table**: ManhattanFireCompanyStreets
   - **Output Table**: FireCompanyTypeStreetLength
   - **Field**: Street_Length
   - **Statistic Type**: Sum
   - **Case Field**: FireBN_Co_Type

3. Close the **ManhattanFireCompanyStreets** table.

4. Open the **FireCompanyTypeStreetLength** table.

In the field **SUM_Street_Length**, you will see the sum (total length) of street segments for each fire company and type.

OBJECTID_1 *	FireBN_Co_Type	FREQUENCY	SUM_Street_Length
1	1	1_1_L	139	35732.046198
2	2	1_10_E	544	96794.841376
3	3	1_10_L	315	74371.810622
4	4	1_15_L	532	91798.225658
5	5	1_4_E	188	35341.360738

> **YOUR TURN**
>
> Join the **FireCompanyTypeStreetLength** table to the **ManhattanFireCompanies** attribute table using **FireBN_Co_Type**. Open and sort **SUM_Street_Length** in descending order. Select the record with the longest street length. Select the next few records to see the location of those fire companies. Close the table, clear all selections, and save your project.

Tutorial 6-6: Using Union on neighborhoods and land-use features

The **Union** tool overlies the geometry and attributes of two input polygon layers to generate a new output polygon layer. In this tutorial, you will use the **Union** tool to combine New York City's Brooklyn borough neighborhoods and land-use polygons to calculate the total land area of each type by neighborhood. The output of a union is a new feature layer of smaller polygons, each with combined boundaries and attributes of both neighborhoods and land-use polygons. You can then calculate land-use type (for example, residential) in each neighborhood. Such information would be useful for an urban planner or sustainability director who is interested in learning about house and land-use development in neighborhoods.

Open tables to study attributes

1. Open **Tutorial6-6.aprx** from the **Chapter6\Tutorials** folder and save the project as **Tutorial6-6YourName.aprx**.

The map contains Brooklyn neighborhood polygons and land use with zoning types. The features clearly do not share the same boundary, and you will see neighborhoods with mixed zoning types.

2. Open the **BrooklynNeighborhoods** attribute table.

   There are 51 neighborhood polygons. The table contains no data about land use in the feature class, but it does contain data about population, housing units, and acres for each neighborhood.

3. Open the **BrooklynLandUse** attribute table.

   There are 1,294 land-use polygons. Fields of interest are the land use for each polygon. Both tables include the length and area for every polygon.

4. Close the attribute tables.

Use Union on features

1. Search for and open the **Union** tool. Apply these settings and run the tool:
   - **Input Features**: BrooklynNeighborhoods and BrooklynLandUse
   - **Output Feature Class**: **BrooklynNeighborhoodsLandUse**

   The result is a new layer of smaller polygons with combined neighborhood and land-use data. Area fields such as **Acres** are not yet calculated for the new, smaller polygons. Each polygon in a neighborhood will have the same acreage.

2. Turn off the original **BrooklynNeighborhoods** and **BrooklynLandUse** layers.

3. Symbolize **BrooklynNeighborhoodsLandUse** using a **No color** fill, an **Outline color** of black, and an **Outline width** of **2**.

4. Zoom to the **BrooklynZoomed** bookmark, click one of the new polygons, and examine the values for neighborhoods and land use.

5. Close the pop-up window.

Calculate acreage

The acres field of each polygon created by the union needs to be updated before you can summarize the land use for neighborhoods.

1. Search for and open the **Calculate Geometry Attributes** tool. Apply these settings and run the tool:
   - **Input Features**: BrooklynNeighborhoodsLandUse
   - **Field**: Acres
   - **Property**: Area
   - **Area Unit**: US Survey Acres

Select and summarize residential land-use areas for neighborhoods

1. In the **Contents** pane, right-click **BrooklynNeighborhoodsLandUse** and click **Properties**.

2. Click the **Definition Query** tab and create the query: **Where LANDUSE2 is equal to Residential**.

3. In the **BrooklynNeighborhoodsLandUse** attribute table, right-click **NTAName**, click **Summarize**, and apply these settings:
   - **Input Table**: BrooklynNeighborhoodsLandUse
   - **Output Table**: BrooklynNeighborhoodsResidentialLandUse
   - **Field**: Acres
   - **Statistic Type**: Sum
   - **Case Field**: NTAName

4. Click **OK**.

5. Open the **BrooklynNeighborhoodsResidentialLandUse** table and sort **SUM_Acres** in descending order.

OBJECTID *	NTAName	FREQUENCY	SUM_Acres
1	21	East New York	39	1490.736881
2	12	Canarsie	39	1399.279915
3	2	Bay Ridge	74	1283.271057
4	44	Sheepshead Bay-...	15	1283.086147
5	28	Georgetown-Mari...	13	1186.707701

The list shows neighborhood names with the highest to lowest residential land use.

6. Using the **NTAName** field, join the **BrooklynNeighborhoodsResidentialLandUse** table to **BrooklynNeighborhoods**.

7. Open the table and move fields **H_UNITS** and **SUM_Acres** to the right of the **NTAName** field.

8. Sort **H_Units** in descending order.

   A planner can now compare the neighborhoods with the highest number of housing units and the number of residential acres.

NTAName *	H_UNITS	SUM_Acres
Crown Heights North	44583	1012.610061
Bay Ridge	39069	1283.271057
Flatbush	38816	995.756047
Bensonhurst West	33939	972.761129
East New York	33772	1490.736881

9. Close the attribute tables and save your project.

Tutorial 6-7: Using the Tabulate Intersection tool

Previous tutorials in this chapter used the **Intersect** and **Union** tools to create feature classes with combined features and data. With these tools, the data (for example, number of housing units or population) is not apportioned (split into parts and allocated) to the new features. For example, if a single neighborhood crosses more than one land-use zone or more than one fire company, the neighborhood housing data should be split between polygons.

In this tutorial, you will use the **Tabulate Intersection** tool to estimate the number of persons with disabilities in fire company boundaries using census tracts and fire company polygons. By default, this tool makes apportionments proportional to the areas of split parts of polygons, such as block groups; this method assumes that the populations of interest are uniformly distributed by area within polygons.

Study tracts and fire company polygons

1. Open **Tutorial6-7.aprx** from the **Chapter6\Tutorials** folder and save the project as **Tutorial6-7YourName.aprx**.

    The map contains Manhattan census tracts classified with the number of persons with disabilities (all disabilities) and Manhattan fire companies.

2. Zoom to the **LowerManhattan** bookmark and turn the **ManhattanFireCompanies** layer on and off to see tracts compared with the borders of fire companies.

    Tracts and fire companies clearly don't share exact borders.

3. Close the **Disabled Person Fire Company Study (Manhattan)** map and open the **Disabled Person Fire Company Study (Upper West Side)** map.

    This map uses a small study area composed of four fire companies and 20 census tracts on the Upper West Side, which have been clipped to better see the workings of the **Tabulate Intersection** tool. Black labels indicate fire companies, white halo labels indicate tract IDs, and yellow halo labels indicate the number of persons with disabilities per tract.

4. Zoom to fire company **76**.

    Five tracts (selected in the figure) intersect the fire company **76** polygon. Four tracts (019300, 018900, 019500, and 019100) and their populations are completely within the polygon for fire company **76**, and one tract (018700) is split between fire companies **76** and **22**. Responsibility for that tract's population of persons with disabilities (880) should be split approximately 50/50, with 440 persons apportioned to fire company **76** and 440 persons apportioned to fire company **22**.

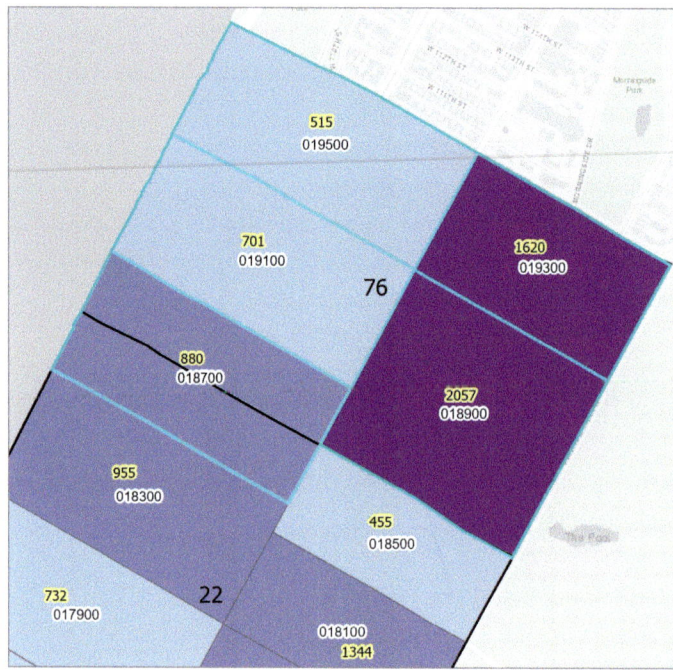

YOUR TURN

Study additional tracts in the other fire company polygons and identify tracts that are split between the fire company polygons.

Use Tabulate Intersection to apportion the population of persons with disabilities to fire companies

1. Search for and open the **Tabulate Intersection** tool. Apply these settings and run the tool:
   - **Input Zone Features**: UpperWestSideFireCompanies
   - **Zone Fields**: FireCoNum
   - **Input Class Features**: UpperWestSideTracts
   - **Output Table**: **DisabledPersonsPerFireCompany**
   - **Class Fields**: TRACT_ID
   - **Sum Fields**: DISABILITY

2. Open the **DisabledPersonsPerFireCompany** table and sort **TRACT_ID** in ascending order.

3. Scroll down and select the two records for tract **018700**.

   The **DISABILITY** field shows that the population of persons with disabilities in the tract is indeed split 50/50 between fire companies **22** and **76**, each with 440 persons. Census tract 019300 (completely within fire company **76**) retains its full population of 1,620 persons with disabilities.

OBJECTID *	FireCoNum	TRACT_ID ▲	DISABILITY
22 8	22	018700	440
23 24	76	018700	440
24 9	22	018900	2
25 25	76	018900	2055
26 26	76	019100	701
27 27	76	019300	1620

   > **YOUR TURN**
   >
   > Clear the selected features. Use the **Summary Statistics** tool to calculate the total number of persons with disabilities in each fire company. Name the **Output Table** as **TotalDisabledPersonsPerFireCompany**. The **Field** will be **DISABILITY** and the **Case Field** will be **FireCoNum**. The results of the total number of persons could be used for planning purposes. Open and review the table. Save your project.

Assignments

This chapter has assignments to complete that you can download with data from ArcGIS Online at links.esri.com/GISTforPro3.4Assignments.

CHAPTER 7

Digitizing

LEARNING GOALS

- Edit, create, and delete polygon features.
- Create and digitize point features.
- Use cartography tools to smooth features.
- Work with CAD drawings.
- Spatially adjust features.

Introduction

Many GIS tools are available to edit and create GIS features. This chapter introduces a few of these tools for manual digitization by tracing. Using basemaps on the computer screen, you will learn how to edit and create vector (point, line, and polygon) map features. You will learn how to create new feature classes and use basemaps or existing layers, such as **Streets**, as spatial guides for digitizing features. Technology such as lidar is also used as a reference for heads-up digitizing. You will learn more about lidar data in chapter 11.

You can use GPS receivers or apps such as ArcGIS Field Maps, which collect longitude and latitude data, to create vector features. CAD and building information modeling (BIM) files can be imported into GIS maps to create feature classes.

In this chapter, you will edit existing features and create features for a rapidly expanding university campus, Carnegie Mellon University (CMU), in Pittsburgh, Pennsylvania. New structures, additions, and renovations to existing buildings are part of a campus master plan. Changes also include new or modified streets, sidewalks, parking lots, and so on. Architects, engineers, and planners need updated GIS features for the campus, and tutorials in this chapter teach the skills to create and edit features in GIS.

Tutorial 7-1: Editing polygon features

In this tutorial, you will move and rotate existing buildings in a layer. You will add vertex points and split polygons to further edit them to match buildings on the **World Imagery** basemap. Imagery maps may change over time, so the images may be slightly different from those in the tutorial.

Move features

1. Open **Tutorial7-1.aprx** from the **Chapter7\Tutorials** folder and save the project as **Tutorial7-1YourName.aprx**.

   The map shows CMU's main campus with buildings as semitransparent features on the **World Imagery** basemap. Because you will make permanent edits to the features, a clean copy of CMU's building polygon features named **Bldgs_Original** is located in **Chapter7.gdb** if you need to start over.

2. Zoom to the **Main Campus** bookmark to examine CMU's main campus and academic buildings.

   A few buildings on the campus are not in their correct locations. You will move them to their correct locations using the **World Imagery** basemap as a reference.

3. Zoom to the **Cohon University Center** bookmark.

4. Using the **Select** tool, click the **Cohon University Center** building polygon.

5. On the **Edit** tab, in the **Tools** group, click the **Move** button.

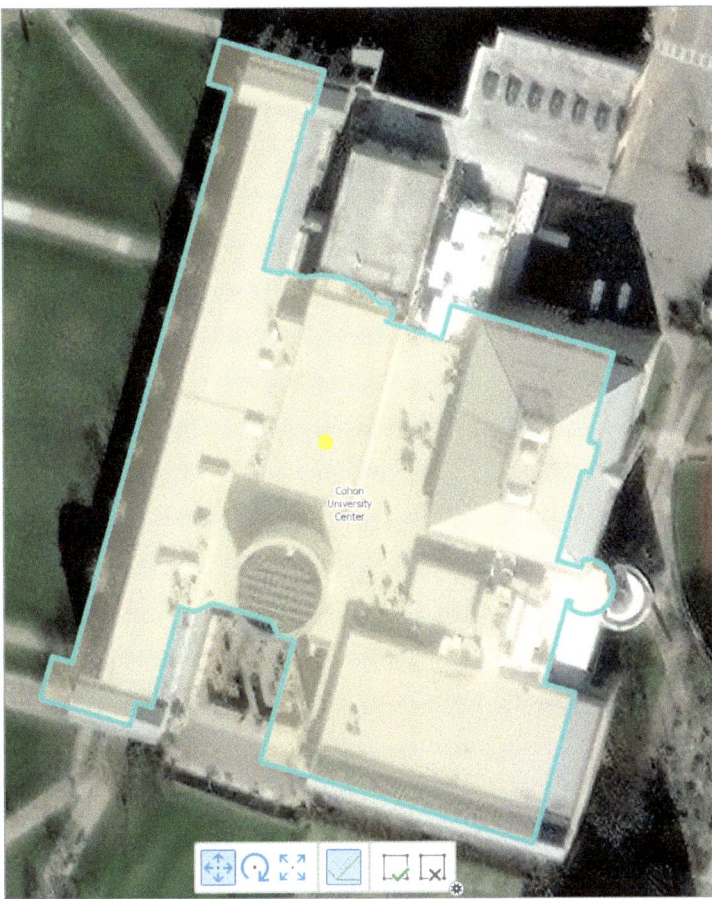

By default, the **Construction** toolbar appears on the active map or when you right-click the map. A yellow dot appears in the center of the polygon indicating that the polygon is ready to edit.

6. Drag the building polygon to match the building outline. Click the **Finish** button.

7. Click anywhere outside the polygon to deselect it.

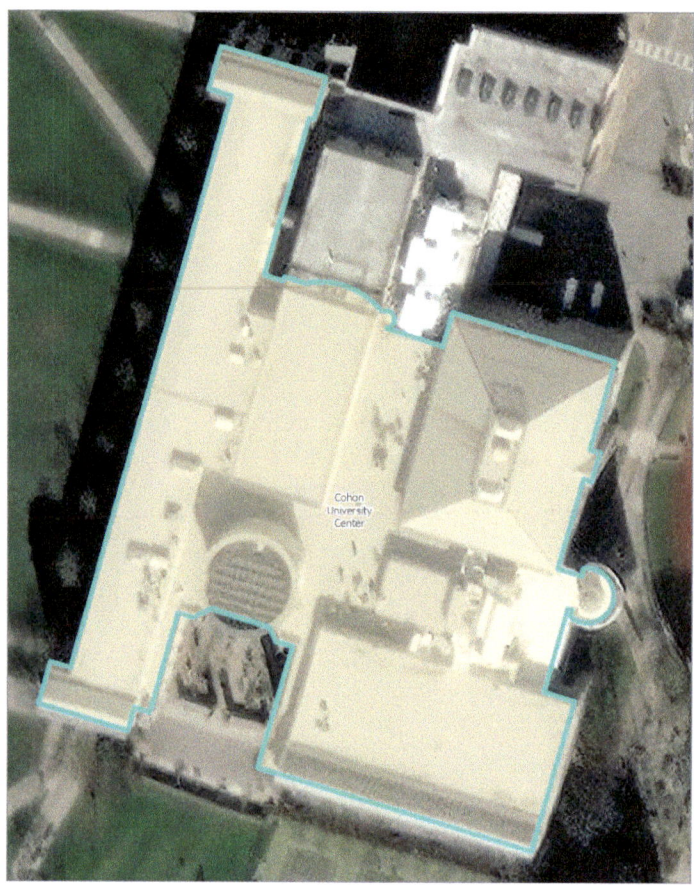

8. On the **Edit** tab, click **Save**. In the **Save Edits** dialog window, click **Yes** to save your edits.

    Saving edits to features differs from saving ArcGIS Pro projects. If you don't want to save changes, click the **Discard** button.

> **YOUR TURN**
>
> Zoom to the **College of Fine Arts** bookmark and move the building polygon to its correct location. Pan the main campus map to find other buildings to move to the correct locations. Save your edits and save the project.

Rotate features

Features may be in the correct location but must be rotated. This applies to CMU's Gates Center for Computer Science.

1. Zoom to the **Gates** bookmark and select the **Gates Center** building polygon.

2. On the **Edit** tab, click **Move**.

   Clicking **Move** enables edits for move, rotate, and scale.

3. In the **Modify Features** pane, click the **Rotate** button.

   A yellow dot and green circle appear in the center of the building, indicating that the polygon is ready for editing.

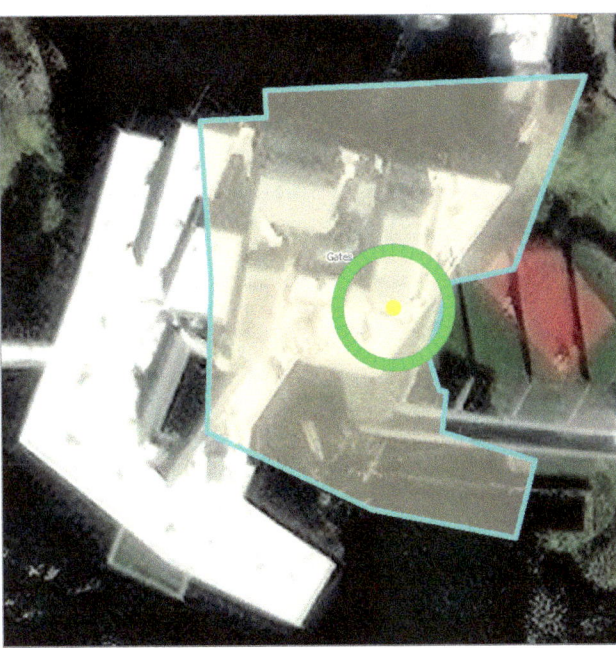

4. Move the pointer to the green circle and drag it to rotate the highlighted outline. Match the approximate orientation. Use the **Move** tool if necessary to position the polygon over the building in the satellite image. Click the **Finish** button.

The building may not align perfectly with the satellite image. You will learn later how to edit vertex points to fix an imperfect alignment.

5. Save your edits.

Add and move vertex points

Features are sometimes represented as simplified outlines, but you can add and modify vertex points to reflect the true building shape. The College of Fine Arts building was drawn as a rectangle, but the building has a U-shaped roof. Adding and editing vertex points will fix this problem.

1. Zoom to the **College of Fine Arts** bookmark. Zoom and pan the map to see the entire building.

2. Select the **College of Fine Arts** building, and on the **Edit** tab, under **Tools**, click the **Edit Vertices** button.

   Four vertex points appear at the corners of the polygon. You must add four more vertices to make the U shape.

3. On the **Construction** toolbar, click the **Add** button, and click four points on the left (western) line of the **College of Fine Arts** building to add four vertex points.

4. Click the **Select By Lasso** button and select the second of the new vertex points.

5. Drag the vertex point to the inside corner of the building.

6. Drag the third vertex point, making a *U* shape.

7. Align the first and fourth vertex points parallel to the points you just moved in the previous steps, outlining the building.

8. Click the **Finish** button and save your edits.

Split features

Two buildings on the campus are drawn as one polygon because they are connected. However, university architects and facility planners must categorize the buildings as separate polygons, each with a separate record in the attribute table. The **Split** tool will accomplish this task.

1. Zoom to the **Baker Porter** bookmark and select the building polygon.

   Baker Hall is the easternmost section of the building, and Porter Hall connects at the western end of the third wing, as shown on the left in the image. This location where the buildings connect is where you will split the polygon.

2. On the **Edit** tab, click the **Split** button.

3. Zoom in. Click outside the polygon on the north side of the small, indented area and double-click the location on the south side where you want the split to occur.

   Refer to the image to see the location of the split.

   This action splits the building into two polygons at this location. There will be a new line in between the two halls.

4. Save your edits.

5. Select the eastern **Baker Hall** polygon. In the **Contents** pane, open the **Bldgs** attribute table.

6. Click the **Show Selected Records** button in the attribute table.

7. In the attribute table, for **NUMBER**, type **2A**. For **BL_ID**, type **BH**. For **NAME**, type **Baker Hall**.

8. Select the western **Porter Hall** polygon. In the attribute table, for **NUMBER**, type **2B**. For **BL_ID**, type **PH**. For **NAME**, type **Porter Hall**.

9. In the attribute table, click **Clear** to clear the selected records.

10. Save your edits, close the attribute table, and save your project.

    Two separate buildings are now labeled for Baker Hall and Porter Hall.

Tutorial 7-2: Creating and deleting polygon features

Campus planners need polygons of open parking lots for a permeable surface and transportation engineering study. A new feature class will highlight these areas for work in a GIS research project related to climate studies and improving campus infrastructure. Point and line feature classes and features can be created using similar steps.

Create a polygon feature class and add a field

1. Open **Tutorial7-2.aprx** from the **Chapter7\Tutorials** folder and save the project as **Tutorial7-2YourName.aprx**.

   The map shows CMU's main campus, existing buildings, streets, and a main campus study area polygon layer.

2. Zoom to the **TepperQuad** bookmark and zoom in to better see the parking lot adjacent to the quad.

   You can create feature classes directly in the **Catalog** pane and add attributes or use the **Create Feature Class** tool and add attributes later in the attribute table.

3. Search for and open the **Create Feature Class** tool. Apply these settings and run the tool:
   - **Feature Class Location**: Chapter7.gdb
   - **Feature Class Name**: **ParkingLots**
   - **Geometry Type**: Polygon

- **Coordinate System**:
  NAD_1983_StatePlane_Pennsylvania_South_FIPS_3702_Feet

4. Open the resulting attribute table and use the **Field** view to create a new field called **LOTNAME**, with the **Alias** as **Parking Lot Name**, **Data Type** as **Text**, and **Length** as **75**.

5. Save your edits and close the **Field** view.

   The new **ParkingLots** feature class is created in the file geodatabase. The feature class is empty now. Later, as you digitize features, polygons will be added to the feature class and become visible on the map.

Add a feature class and create polygons

Before you start adding features, you will change the style for the layer's symbol.

1. In the **Contents** pane, change the **Color** of the **ParkingLots** layer to a dark red, a **Transparency** of **25%**, with a white **Outline color**, and an **Outline width** of **2** pt.

2. On the **Edit** tab, in the **Features** group, click **Create**.

3. In the **Create Features** pane, click **ParkingLots** and confirm that **Polygon** is selected.

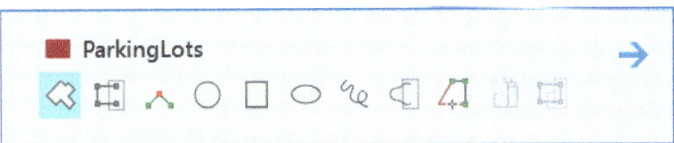

4. On the **Configure** toolbar, with the **Line** button active, click to add points and draw a feature outlining the parking lot, similar to the figure. Double-click the last vertex point to finish the polygon. Click the **Finish** button.

    If the imagery map has an updated parking lot, use that as a guide to digitize the lot.

5. In the **ParkingLots** attribute table, for **Parking Lot Name**, type **TepperQuad**.

6. Save your edits and clear the selected polygon.

> ### YOUR TURN
> Create a point feature class called **BusStops** using the same coordinate system as **ParkingLots**. Turn off all layers, change the basemap to **Streets**, and digitize the approximate locations near the **Cohon University Center** for the following bus stops in the image. In the symbol **Gallery**, search for the **Bus** symbol, and use it as the point symbol. Save your edits and the project.
>
>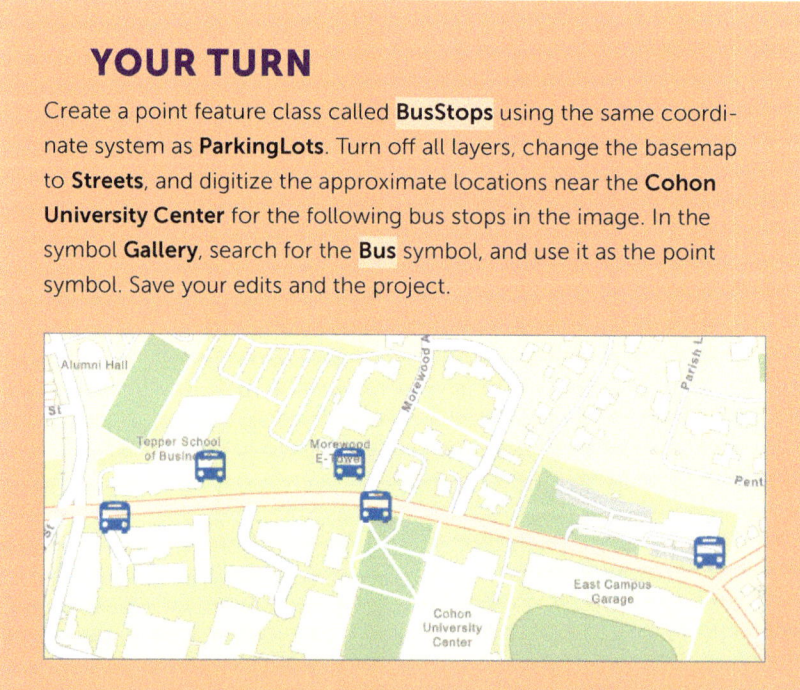

Delete polygons

Campus development often includes the demolition of existing buildings. As part of the university's expansion, four buildings will be demolished and must be deleted from the GIS buildings layer.

1. Zoom to the **Arts Park** bookmark.

2. Turn on the **Streets** and **Bldgs** (buildings) layers and change the basemap to **Imagery**.

3. Hold the **Shift** key and select the four buildings to the east.

4. On the **Edit** tab, click **Delete**.

5. Save your edits.

Use the Trace tool to create a polygon feature

Campus architects and campus planners want to know the square footage of a study area of the main campus where most of the academic buildings are located. Planners identify six streets (**Forbes**, **Boundary**, **Schenley**, **Frew**, **Tech**, and **Margaret Morrison**) that are used to create the study area. The **Trace** tool creates a polygon using these streets as guides.

1. Zoom to the **Main Campus** bookmark and turn off all layers except **MainCampusStudyArea**, **Streets**, and the basemap.

2. On the **Edit** tab, click **Create** and enable **Snapping**.

3. In the **Create Features** pane, click the **MainCampusStudyArea** layer and the **Trace** button.

4. Click the intersection of **Boundary St.** and **Forbes Ave.** Trace the pointer south along **Boundary St.**, east on **Schenley Dr.**, north and east on **Frew St.**, north and east on **Tech St.**, northeast on **Margaret Morrison St.**, and northwest on **Forbes**.

5. Double-click the original point at the intersection of **Boundary St.** and **Forbes Ave.** Click the **Finish** button.

   The new study area polygon matches the existing street centerlines exactly.

6. Save your edits and clear the selection.

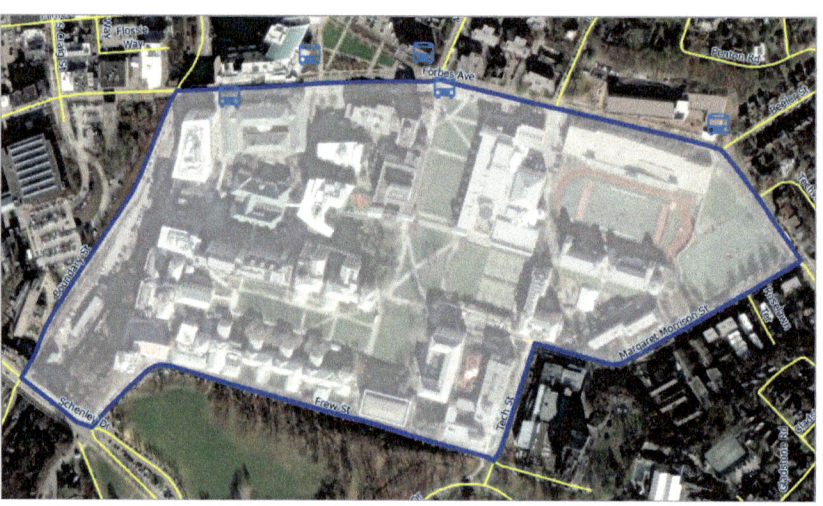

7. Open the **MainCampusStudyArea** table and note the **Shape_Length** and **Shape_Area** values, which are in feet.

   It's useful for campus architects and planners to know the acreage of the campus. This could be automatically calculated for the new **MainCampusStudyArea** polygon using the **Calculate Geometry Attributes** tool covered in chapter 6.

8. Save your project.

Tutorial 7-3: Using cartography tools

In addition to using editing tools, you can modify GIS features using cartography tools. A useful tool to improve the aesthetic or cartographic quality of polygons is the **Smooth Polygon** tool. Organizations such as the US Census Bureau sometimes digitize lines and polygons at a small scale. The features are digitized with just a few line segments and don't match the true geography. Smoothing features can fix this problem.

Open the Tutorial 7-3 project

1. Open **Tutorial7-3.aprx** from the **Chapter7\Tutorials** folder and save the project as **Tutorial7-3YourName.aprx**.

2. Zoom to the **Flagstaff Hill & Panther Hollow Lake** bookmark.

   This tutorial contains a map with semitransparent polygon features, a park and golf course adjacent to the university's campus, and a lake south of campus. The map is zoomed to the park and lake.

Smooth a green space polygon

Flagstaff Hill is part of Pittsburgh's Schenley Park between CMU's campus and Phipps Conservatory. The Flagstaff Hill polygon was roughly drawn with just a few line segments and can be smoothed.

1. Search for and open the **Smooth Polygon** tool. Apply these settings and run the tool:
   - **Input Features**: Greenspaces
   - **Output Features**: GreenspacesSmoothed
   - **Smoothing Algorithm**: Polynomial Approximation with Exponential Kernel (PAEK)
   - **Smoothing Tolerance**: 150 US Survey Feet

   Smoothing Tolerance is important when creating the vertices of the new path. A shorter length will result in a more detailed (or smoother) path but will take longer to process.

2. Turn off the **Greenspaces** and **Water** layers and zoom to the **GreenspacesSmoothed** layer.

   The result is a new feature class of smoothed polygons for Flagstaff Hill and the polygons for Schenley Golf Course.

> **YOUR TURN**
>
> Turn on the **Water** layer and zoom to it. Use the **Smooth Polygon** tool to smooth the water feature of **Panther Hollow Lake**. Save the new feature as **WaterSmoothed** to **Chapter7.gdb**. Turn off the original **Water** layer and save your project.

Tutorial 7-4: Transforming features

Administrators, architects, and facility planners often combine GIS map layers and computer-aided design (CAD) drawings and building information models, or BIM models, of separate buildings for strategic planning, including understanding the occupancy and use of every space on campus. CAD drawings and BIM models, however, use Cartesian coordinates (for example, the lower-left XY position of most CAD and BIM files is 0,0) and are not geographically referenced to geographic coordinate system, state plane, or Universal Transverse Mercator (UTM) coordinates. CAD drawing units are also different from GIS units. For example, the leading CAD software, AutoCAD, uses one inch or one millimeter as the unit, and GIS maps generally use feet or meters as the unit. Transforming features in GIS makes aligning CAD drawings to GIS maps easy, regardless of drawing or map coordinates and units. In this tutorial, you will import one campus building, Hamburg Hall. You'll also import a CAD drawing, assign a map projection, and use the **Georeference** tool to locate it at approximately the correct place on CMU's campus. Then you'll use the **Transform** tool to place the building more precisely at the correct geographic location on a campus map.

Add a CAD drawing and view layers

1. Open **Tutorial7-4.aprx** from the **Chapter7\Tutorials** folder and save the project as **Tutorial7-4YourName.aprx**.

   The map shows CMU's campus with building outlines, a map projection of NAD 1883 StatePlane Pennsylvania South, and map and display units in US feet.

   Next, you will add an AutoCAD drawing that includes polygons for every space on the first floor of CMU's Hamburg Hall academic building. If you checked the spatial reference in **Layer Properties**, you would see it says Unknown Coordinate System.

2. Click **Add Data**, browse to **Chapter7**, click **Data**, double-click **HBH1.dwg**, click **Polygon**, and click **OK**.

3. Zoom to the **HBH1-Polygon** layer.

   The drawing unit is one inch, with the building in the CAD drawing appearing as 12 times its actual size in the **StudyAreaBldgs** layer. The CAD drawing is also located south and west of the real building.

   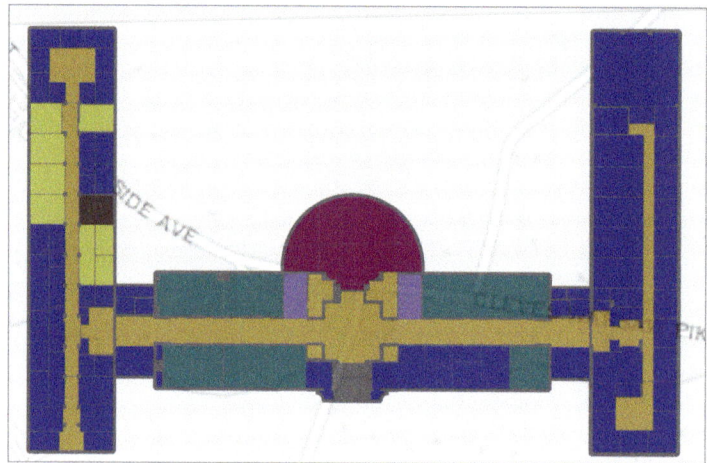

4. In the **Contents** pane, click the arrow next to **HBH1-Polygon** to expand the layers.

   CAD drawings contain layers in one drawing and are color-coded according to the layer color assigned (shown as a number) in the CAD drawing. Also listed are the CAD line type (**Continuous**). You cannot edit CAD drawings directly, so you will export the drawing to a feature class after you georeference it to its approximate campus location.

   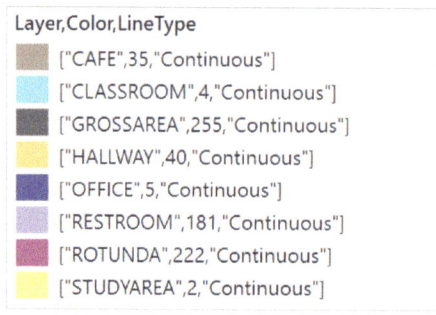

Use Georeference to move and rotate the CAD drawing

Here, you use **Georeference** tool and two control points to move and rotate the CAD drawing from its arbitrary location to a better location on the campus map.

1. In **Contents**, click the **HBH1-Polygon** layer.

2. On the **CAD Data** tab, in the **Alignment** group, click **Georeference**.

3. In the **Adjust** group, click **Add Control Points**.

4. Click the lower-left corner of the building.

   This will establish the first control point. You will next click the same location on the building on the campus map.

   Tip: The **StudyAreaBldgs** layer faces the opposite direction.

5. In the **Contents** pane, right-click **StudyAreaBldgs** and click **Zoom To Layer**.

6. Click the corresponding point on the building outline.

   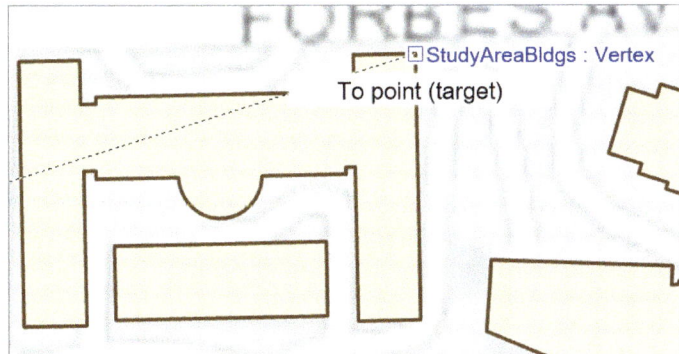

7. Right-click **HBH1-Polygon** and zoom to the layer.

8. Click the lower-right corner of the building.

9. Right-click **StudyAreaBldgs** and zoom to the layer.

10. Click the corresponding point on the building outline.

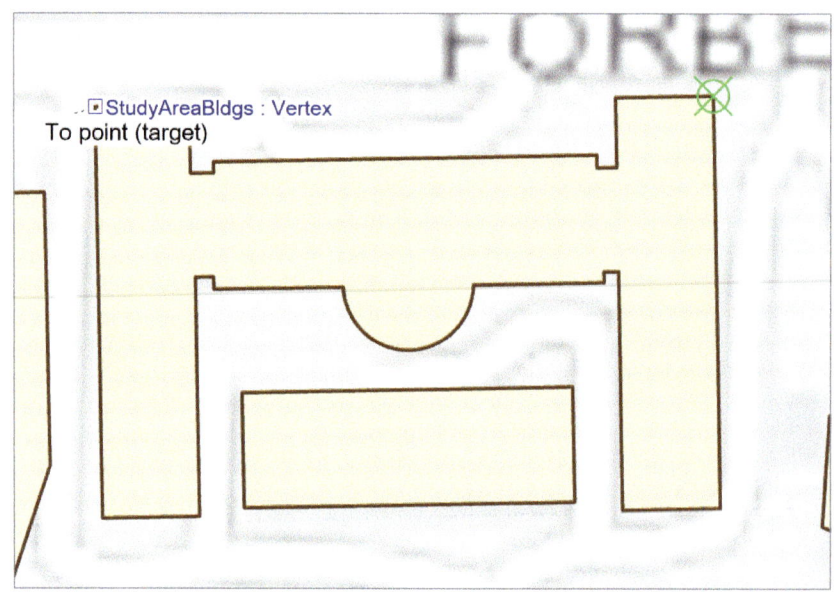

11. In the **Adjust** group, click **Apply**, and in the **Save** group, click **Save** > **Yes**.

12. Close the **Georeference** tab.

   The CAD drawing will now be in the approximate location of the building.

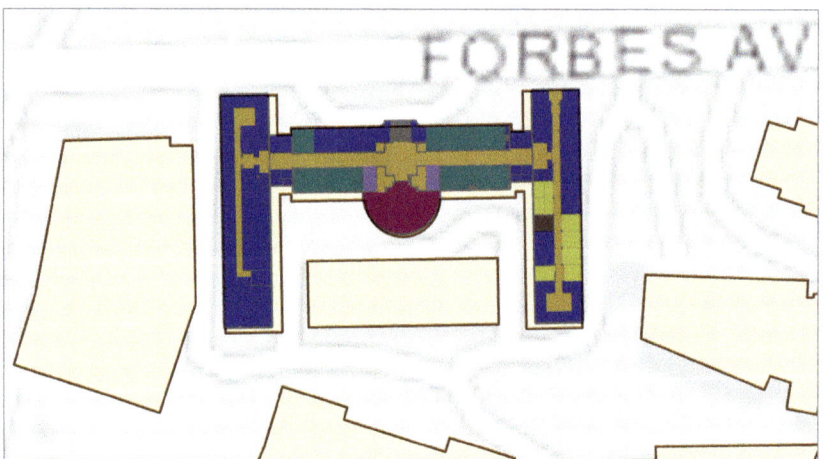

Export the CAD file and assign a projection

The results of exporting a CAD drawing are that properties of the drawing are added as fields in the attribute table and the drawing can more easily be edited using GIS tools.

1. In the **Contents** pane, right-click **HBH1-Polygon** and click **Data > Export Features**.

2. In the **Export Features** tool, for **Output Feature Class**, save as **HBH1SpacePlan** in **Chapter7.gdb**.

3. Click the **Environments** tab. For **Output Coordinate System**, select **StudyAreaBldgs**.

   This will automatically assign **NAD_1983_StatePlane_Pennsylvania_South_FIPS_3702_Feet** as the projection for the new feature class.

4. Click **OK**.

   HBH1SpacePlan is automatically added to the map in the new location. You will transform the new layer to its more precise location on the campus map.

5. Remove the **HBH1-Polygon** CAD drawing.

Explore attributes and classify layers

Before transforming the drawing, you will explore the attributes of the exported CAD drawing and classify spaces based on their layer assignment.

1. Open the **HBH1SpacePlan** attribute table.

   The **Layer** field shows designations of space use. Every polyline in AutoCAD became a polygon in the new feature class, and the layer on which the CAD polyline resided became the layer designation (for example, **Office**, **Hallway**, **Classroom**, and so on).

2. Close the attribute table.

3. Search for and open the **Apply Symbology From Layer** tool. Apply these settings and run the tool:
   - **Input Layer**: HBH1_SpacePlan
   - **Symbology Layer**: Browse Chapter7\Data\SpacePlan.lyrx

The CAD drawing will now nicely show the spaces of each room type and a better legend.

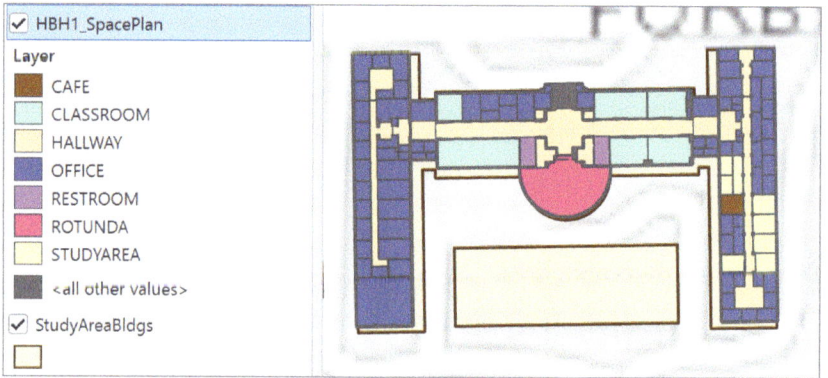

Transform polygons

Depending on the purpose of your project, the current location may be "close enough," but the **Transform tool** will move the building to be more precise with the study area building outline. You will transform the building's polygons using links that better align the floor-plan polygons with the building on the map.

1. Turn off the **StudyAreaBldgs** layer and select all the **HBH1_SpacePlan** features.

2. Turn the **StudyAreaBldgs** layer on and, with all the **HBH1_SpacePlan** features selected, and on the **Edit** tab, in the **Features** group, click **Modify**.

3. In the **Modify Features** pane, click the **Transform** button.

4. Under **Transformation Method**, select **Similarity 2D** and click **Add new links**.

5. Click the corners of the **HBH1_SpacePlan** layer and the corresponding corners of the **Hamburg Hall** building on the campus map as shown. Use your wheel button to zoom in and out while adding points, clicking and holding the wheel button down to pan.

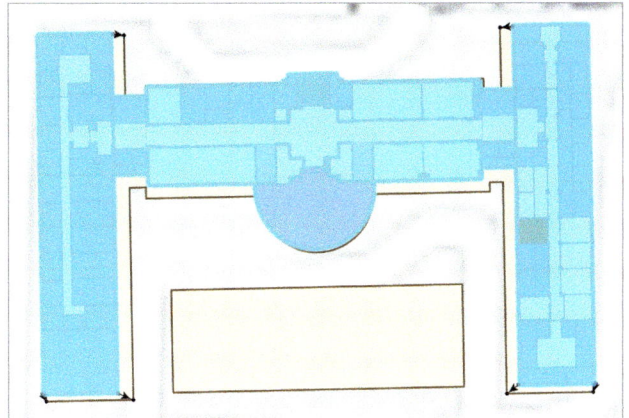

6. In the **Modify Features** pane, click the **Transform** button on the bottom right.

7. Wait while the polygons are moved on the map to the more precise location.

8. Save your edits and clear selected features.

9. Zoom to the **StudyAreaBldgs** layer to better see the result.

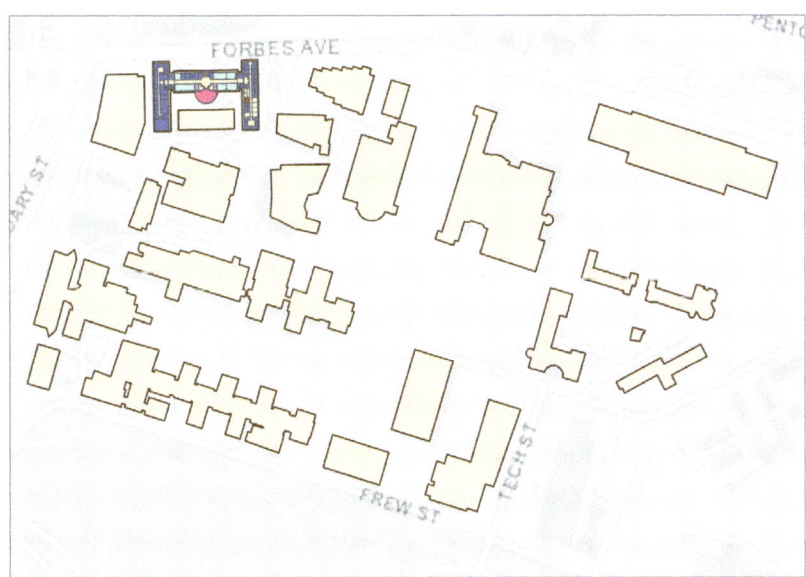

10. Save your project.

Assignments

This chapter has assignments to complete that you can download with data from ArcGIS Online at links.esri.com/GISTforPro3.4Assignments.

CHAPTER 8

Geocoding

LEARNING GOALS

- Learn about the geocoding process.
- Geocode using zip codes.
- Geocode addresses using streets.

Introduction

Geocoding is a GIS process that matches location fields in tabular data, such as 123 Main Street, Pittsburgh, PA 15213, to corresponding fields in existing feature classes, such as TIGER/Line Streets, to map the tabular data. An example is the survey data from this chapter that lists street addresses and zip codes of the people surveyed at an art show. Using these location fields, you can geocode their locations using zip code polygon or street feature classes. Another example is transaction data collected by organizations. Because transaction data records, such as for deliveries of appliances, often include location fields, it's possible and useful to map these locations—for example, to optimize routing of delivery trucks that make several stops. Another example of transactions is patients in a hospital. In this case, it's useful to map the residences of patients to identify the service area of the hospital.

The source data that you will geocode in the beginning of this chapter is from a survey taken by an arts organization of attendees of its annual art show. The arts organization wants to analyze locations of attendees to better target future marketing efforts. You'll geocode first for a multistate region by zip code and then for the home county of the arts organization by street address. Many sources of data, including surveys, often include only zip code data and not street addresses. The survey data of this chapter, however, has both for comparison.

One problem with geocoding is that source data suppliers (for example, survey respondents for the survey data) and data entry workers can write or type anything they want for an address, including misspellings, abbreviations, omissions, and place-names such as "University of Pittsburgh" instead of an address. Consequently, exact matching of sources to reference data is not possible. Instead, GIS must use so-called fuzzy matching and make matches that are approximate instead of always 100 percent accurate. For example, the address "123 Fleet St." may be on a reference street map, but a surveyed person may have written "234 Fleat" for source data, with a misspelling of "Fleet" and without the "St." besides the misnumbering of the address. A fuzzy-matching algorithm nevertheless may determine that the source address is close enough to the reference street address and plot the residence at 234 Fleet St., which may be close to the original address of 123 Fleet St.

A rule-based expert system is software that makes fuzzy matches, and the geocoding software in ArcGIS Pro is such a system. The system attempts to use the thought processes and rules that an expert would use to accomplish a complex and ambiguous task. For example, you can think of the geocoding expert system as attempting to mimic what a resourceful mail delivery person would do, using their expert knowledge to get a badly addressed piece of mail to the right address. The following expert system components are used in ArcGIS Pro:

- Source table, including geocodes (in this chapter, street addresses and zip codes) to be mapped
- Reference data, which has mapped features such as street centerlines or zip code polygons and corresponding fields such as street names that can be fuzzily matched to the source table's geocodes
- Geocoding tool, with the algorithms, rules, and user interface to perform geocoding
- Locator, a reusable set of files that include all geocoding parameter values and data for a specific kind of reference data

To account for spelling errors, an algorithm computes a Soundex key, which is a code assigned to names that sound alike (for example, "Fleet" and "Fleat" both have Soundex key F43) and identifies candidate matches of source and reference street addresses. The algorithm starts with a score of 100 for each source record and subtracts penalty points for each problem encountered. If the end score is greater than the minimum candidate parameter value set by default or by users, a reference location is a candidate for matching. For two or more candidates, the candidate with the maximum score is chosen as the estimated location. If there is a tie, one of the tying locations can be arbitrarily assigned as the match unless users choose to not accept ties.

In this chapter, you will geocode using zip codes and street centerlines. People typically disclose their zip codes in surveys and get them right, so geocoded results are generally complete and accurate, albeit only at the zip code level. Often, a zip code may be the only available data type and will suffice for marketing purposes. Service or product delivery and other location-based needs require more precise locations. Street centerlines are sufficient for many purposes but certainly not for locating in-ground natural gas lines during construction digging. You can geocode with zip codes and street centerlines with free map layers downloaded from the internet (see chapter 5). However, cities and states perform many other kinds of geocoding, often more precisely—for example, using land parcel centroids with street addresses as provided by many city governments. Esri provides the accurate and current ArcGIS Online World Geocoding Service. If you are in a class, however, ask your instructor before using this service because using the service consumes credits that must be purchased from Esri.

Street centerline maps, available from the US Census Bureau's TIGER/Line data and from vendors, are widely used for geocoding street addresses but are limited by having house numbers only on the left and right sides or the beginning and end of each one-block-long street segment. Consequently, addresses within blocks are linearly interpolated (for example, 150 Main St. is plotted halfway along the street segment, with ranges 100 to 198 and 101 to 199 on either side) and are not exact locations.

Source data records are not always matched when geocoding. A performance measure for geocoding is the percentage of source addresses that get matched and plotted using the reference data. To compute match rates, divide the total number of addresses matched by the result of subtracting all records in source data that do not have addresses (location fields that are blank, do not start with a house number, are not street intersections, and so on) from the total number of addresses in the source data:

Total number of addresses matched / (total number of addresses − number of records with no addresses)

Yet there is no way to judge how many matches are truly correct. Organizations that critically depend on geocoding (such as 911 emergency calls for services from police, fire, and ambulance responders) review nonmatches and incorrect matches to improve their maps and procedures for obtaining correct source data from callers.

Tutorial 8-1: Geocoding data using zip codes

In this tutorial, you will geocode survey data collected by a Pittsburgh arts organization that holds an event each year attended by persons across the country but mainly by those residing in the four-state region of Pennsylvania, Ohio, Maryland, and West Virginia. You can use such survey data, if geocoded, for marketing, philanthropy, or other forms of communication with its patrons.

Build a zip code locator

1. Open **Tutorial8-1.aprx** from **Chapter8\Tutorials** and save the project as **Tutorial8-1YourName.aprx**.

   The **AttendeesPARegion.csv** table is the source data for geocoding, and **PARegionZIP** (zip code polygons) is the reference data.

   Recall that a geocoding locator is a set of files that store parameters and other data for the geocoding process.

2. Search for and open the **Create Locator** tool. Apply these settings and run the tool:
   - **Country or Region**: United States
   - **Primary Table(s)**: PARegionZIP
   - **Role**: ZIP (selecting the ZIP field for the locator Role makes this a zip code locator)
   - ***ZIP**: GEOID10
   - **Output Locator**: **PARegionZIP_CreateLocator**
   - **Language Code**: English

3. In the **Catalog** pane, expand **Locators**, right-click **PARegionZIP_CreateLocator**, and click **Properties**. Click the **Geocoding options** tab and expand **Match Options**.

   These options are the default geocoding parameters used for this zip code locator—namely, minimum match score and minimum candidate score. For each matching problem detected, the geocoding algorithm subtracts penalty points from a starting score of 100. When the algorithm finishes scanning for problems, if the match score is 75 or higher and the candidate score is 70 or higher (as set here by default), the minimum score rules are met. You can change the minimum match or minimum candidate threshold scores to adjust (tune) the fuzzy-matching process and make the process stricter or less strict. High values allow fewer match errors, and low values allow more match errors. You will use the defaults for geocoding the arts organization survey data.

4. Close the **Locator Properties** window.

Geocode data by zip code

1. Open the **AttendeesPARegion.csv** table.

The table has 1,123 survey responses, and if you sort by **ZIPCode** and scroll down, you'll see no records with missing zip code values.

2. Close the table.

3. Search for and open the **Geocode Addresses** tool. Apply these settings and run the tool:
   - **Input Table**: AttendeesPARegion.csv
   - **Input Address Locator**: PARegionZIP_CreateLocator

   Do not select **ArcGIS World Geocoding Service** or **Esri World Geocoder for Address Locator** because using those will consume credits from your ArcGIS account.
   - **Input Address Fields**: Single Field
   - **Single Line Input**: ZIPCode
   - **Output Feature Class**: **Attendees**

4. At the lower left of the **Geoprocessing** pane, click **View Details**. View the **Messages** tab to see that 1,120 out of 1,123 (99.73%) attendee records were matched.

   Considering that all input records had zip codes and zip code data is generally accurate, this high match rate is as expected.

5. Close the pane.

6. Symbolize **Attendees** with **Circle 3**, the **Color** red, and **Size 5** pt.

Because each matched source address adds a point at the zip code centroid, many attendee points may be on top of one another for a given zip code area.

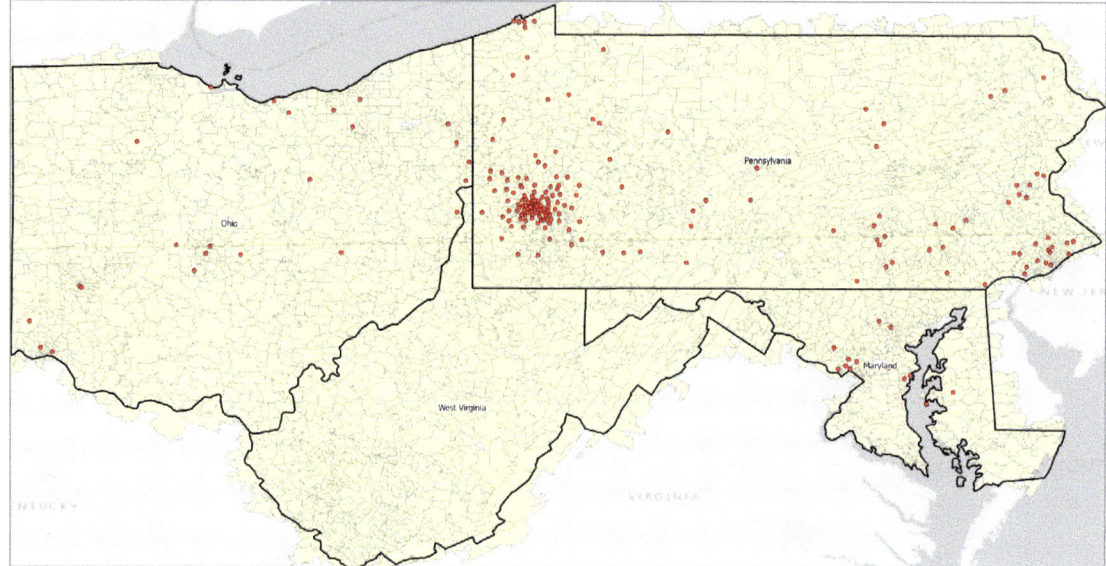

7. In the **Contents** pane, open the **Attendees** attribute table.

8. Sort **Status** in descending order to find the three unmatched records and scroll to the right of the table.

   The three unmatched records have street addresses, cities, and states, which is enough information to look up the correct zip codes at the US Post Office website. The first unmatched record has the incorrect zip code value, **15230**. The value should be 15213. The other two records have correct zip code values, but the reference data, **PARegionZip**, is missing those zip code polygons, indicating another type of error that can occur.

9. Close the table and save your project.

Rematch attendee data by zip code

The match rate, 99.7 percent, is extremely high and well above the threshold for any marketing decision-making or other management purposes, so you can use the Attendees map without any changes. But for practice, you'll rematch to 100 percent. You'll correct the zip code in one record and pick approximate points on the map for the two records that had no reference data.

1. In the **Contents** pane, right-click **Attendees** and click **Data** > **Rematch Addresses**.

   The first unmatched record has the incorrect zip code, **15230**. In the next step, you'll correct the zip code value as part of the rematching process.

2. In the **Rematch Addresses** pane, for **Single Line Input**, type **15213** and click **Apply**.

   The new zip code yields a candidate with a score of **100**, which is above the threshold of 75.

3. Click the **Match** button (green check mark).

   That record is now matched and mapped.

4. Turn off **PARegionZIP**, change the basemap to **Streets**, and zoom in northwest of Pittsburgh.

   Any location will do for the sake of learning how to add a point to a map in the geocoding process.

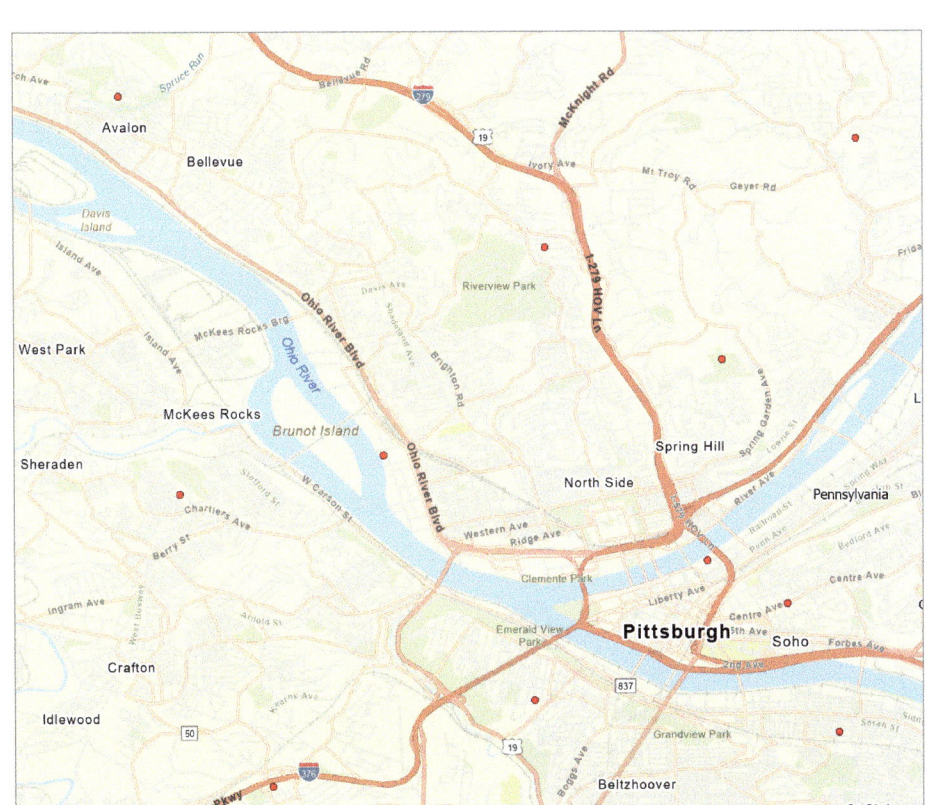

5. In the **Rematch Addresses** pane, click the **Pick from Map** button and click the map at a location of your choice.

6. Click the **Match** button and click **Yes**. The record is matched, and a point is added to the map.

7. In the **Rematch Addresses** pane, click **Save Edits** and click **Yes**.

8. Zoom to the **Region** bookmark.

> **YOUR TURN**
>
> Similarly, match the remaining unmatched record to any location in eastern Pennsylvania. This work is just for practice, so it's not important what location you pick. Save your edits and the project.

Symbolize using the Collect Events tool

Now you have attendees' survey data geocoded to zip code centroids, generally with many attendees at each centroid. To symbolize the attendees, you will count the number of attendees at each zip code and use graduated symbols, with symbol size increasing as the number of attendees increases.

1. Turn off **Attendees**, change the basemap back to **Light Gray Canvas**, and zoom to the **Region** bookmark.

2. Clear any selected records.

3. Search for and open the **Collect Events** tool. Apply these settings and run the tool:
   - **Input Incident Features**: Attendees
   - **Output Weighted Point Feature Class**: **Attendees_CollectEvents**

4. Zoom in to the southwest corner of Pennsylvania where Pittsburgh is located and save your project.

Collect Events has done the hard work of counting by zip code and applying graduated symbols.

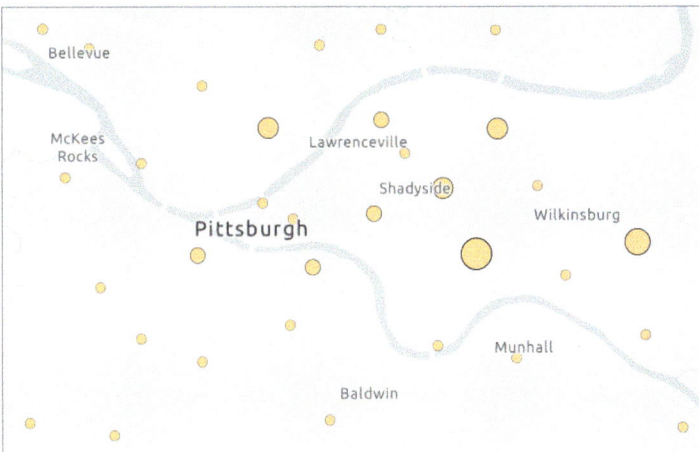

Tutorial 8-2: Geocoding street addresses

This tutorial uses the arts event data from tutorial 8-1 but includes records only for Allegheny County, which includes the city of Pittsburgh, and adds street addresses. Allegheny County is the local market for the arts event, and more detailed location data on attendees is desirable for that market. You will geocode by street address to place unique points on the map for attendees in the county. You'll use the same workflow as you did with zip code matching: Build a locator (this time using street centerlines as the reference data) and geocode the source data of survey respondents. As a step for tuning the geocoding process, however, you will set low values for the minimum match and candidate score parameters of the locator—10 and 10, respectively—replacing the defaults. As a result, geocoding will make obvious errors for low parameter values. By analyzing the attribute table of the geocoded output, you can select values for the minimum match and candidate score parameters that are suitable for the survey data. Scores below those you select will make identifiable match errors, and scores above will be accurate enough for marketing.

Build a street locator and set its geocoding option

1. Open **Tutorial8-2.aprx** from **Chapter8\Tutorials** and save the project as **Tutorial8-2YourName.aprx**.

   You are seeing 92,430 block-long street segments of Allegheny County on your map.

2. Open **AttendeesAlleghenyCounty.csv**.

   The spreadsheet includes street address, city, state abbreviation, and zip code data in four separate fields.

3. Close the table.

   This time, in the **Create Locator** tool pane, you will set its **Primary Table** (reference data) to **Streets** and its **Role** to **Street Address**.

4. Search for and open the **Create Locator** tool. Apply these settings and run the tool (field names starting with an asterisk are required):
   - **Country or Region**: United States
   - **Primary Table(s)**: Streets
   - **Role**: Street Address
   - *****Left House Number From**: LFROMADD
   - *****Left House Number To**: LTOADD
   - *****Right House Number From**: RFROMADD
   - *****Right House Number To**: RTOADD
   - *****Street Name**: FULLNAME
   - **Left ZIP**: ZIPL

   Several Allegheny County cities may have streets with the same address numbers and street name, such as "100 Main Street." Because street names are unique in zip code areas and cities, including the zip code in the matching process guarantees that you will create a match at the correct "100 Main Street" location.
   - **Right Zip**: ZIPR
   - **Output Locator**: Streets_CreateLocator
   - **Language Code**: English

5. In the **Catalog** pane, expand **Locators**, right-click **Streets_CreateLocator**, and click **Properties**. Click the **Geocoding options** tab and expand **Match Options**.

6. For **Minimum match score** and for **Minimum candidate score**, type **10** for each and click **OK**.

Geocode attendee data by street address

1. Search for and open the **Geocode Addresses** tool. Apply these settings and run the tool:
    - **Input Table**: AttendeesAlleghenyCounty.csv
    - **Input Address Locator**: Streets_CreateLocator
    - **Input Address Fields**: Multiple Field (some fields fill out automatically)
    - **Address or Place**: Address
    - **City**: City
    - **State**: State
    - **Zip**: ZIP_Code
    - **Output Feature Class**: **AttendeesAlleghenyCounty_Geo1010**

2. After the tool finishes, click **View Details** > **Messages**.

    A total of 918 of 932 records are matched, for a match rate of 98.5 percent. Of the remaining records, 10 are unmatched and 4 are tied. Several of the matched records have incorrect locations, considering the low match minimums you entered for the street address locator. The geocoding algorithm may change with new software releases, so your match rate may be different.

3. Close the pane.

Select minimum candidate and matching scores

Next, you will rearrange the order of fields displayed in the geocoded results attribute table and then sort the table. Then you will compare the input attendee addresses and zip codes versus those matched by geocoding.

1. Open the attribute table for **AttendeesAlleghenyCounty_Geo1010**.

2. In the upper right of the attribute table, click the **Menu** button, uncheck the box for **Show Field Aliases**, and click **Fields View**.

3. Scroll down to the bottom of the **Fields** table, press and hold the **Ctrl** key, and select **USER_Address** and **USER_ZIP_Code**.

    These are input source values from the survey.

4. Drag the selected rows upward and drop them after **Match_addr** near the top of the table.

5. Save and close the **Fields** table.

6. Sort the attribute table by sorting **Score** in ascending order.

7. Compare **Match_addr** (the matched address and zip code in the **Streets** reference feature class) with **USER_Address** and **USER_ZIP_Code**.

As you scroll down through the table, there are obvious errors and some good matches. Some **Match_addr** values don't have street numbers, such as 2nd St, 15045 (Score 79), whereas corresponding **Address** values do, such as 2321 2nd St 15045. In such cases, the **Streets** reference data has some street segments with the correct name and zip code but without street numbers (some **LFROMADD, LFTOADD, RFROMADD,** and **RTOADD** values were missing) in the range needed for a match. The geocoding algorithm picks the center point on one such street segment as an approximation. Such approximations are probably good enough for marketing purposes, but for better-quality matching, a score around 88 or 90 results in a good minimum score. A score in that range would ensure that most, if not all, matching addresses will have matching street numbers.

> **YOUR TURN**
>
> To identify the number of matched records with a minimum score of **90**, use the **Select By Attributes** tool to build a query. You should find 873 (93.7 percent) matched, which is good geocoding performance.

Produce final geocoding results

1. Clear any selected records and close the **AttendeesAlleghenyCounty_Geo1010** attribute table.

2. In the **Contents** pane, right-click **AttendeesAlleghenyCounty_Geo1010**, click **Properties**, click the **Definition Query** tab, and add and apply a new definition query for **Where Score Is Greater Than or Equal to 90**.

3. In the **Contents** pane, turn off the **Streets** layer. Rename **AttendeesAlleghenyCounty_GEO1010** as **AttendeesAlleghenyCounty** and symbolize it with **Circle 3** and **Size 5** pt.

Obvious clusters of attendees are interspersed with gaps with no attendees. This is useful spatial information for marketing the arts event.

4. Save and close your project.

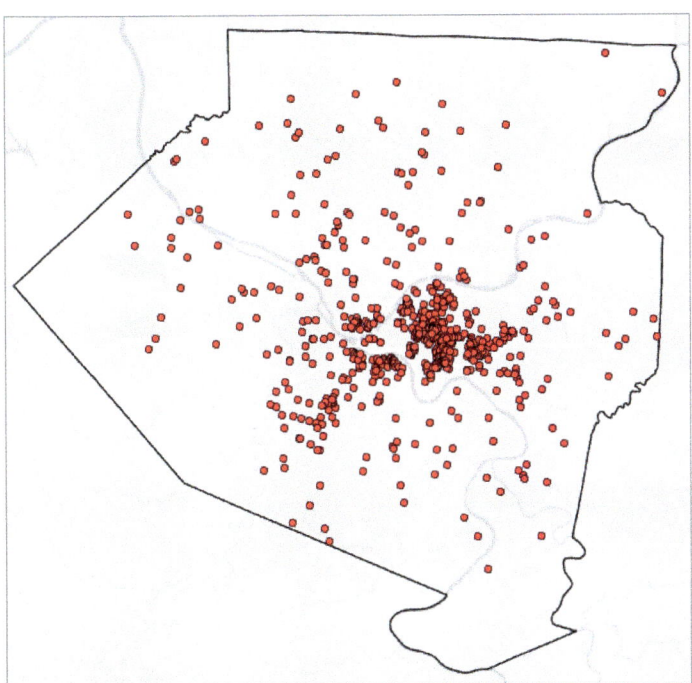

Assignments

This chapter has assignments to complete that you can download with data from ArcGIS Online at links.esri.com/GISTforPro3.4Assignments.

PART 3

Applying advanced GIS technologies

CHAPTER 9

Spatial analysis

LEARNING GOALS

- Use buffers for proximity analysis.
- Use multiple-ring buffers.
- Create service areas of facilities to estimate a gravity model of demand versus distance from the nearest facility.
- Optimally locate facilities using ArcGIS Network Analyst™.
- Perform cluster analysis to explore multidimensional data.

Introduction

Maps sometimes require more than a visualization of spatial data for users to answer questions and solve problems. The right data may be on the map, but analytical methods may be needed to get answers or solutions to problems. This chapter covers four spatial analytical methods: buffers, service areas, facility location models, and clustering. The application areas include building buffers and estimating gravity models for the number of youths using public swimming pools as a function of the distance to the nearest pool from their residences. The analysis suggests which public swimming pools to keep open during a budget crisis. A second application uses cluster analysis to investigate demographic and spatial patterns of arrested persons for serious violent crimes.

This chapter introduces a new spatial data type, the network dataset, which is used for estimating travel distance or time on a street network. ArcGIS Online provides accurate network analysis with network datasets for much of the world. Because these services require using ArcGIS credits for purchase, however, ask your instructor if you have credits available. Otherwise, for instructional purposes, this chapter uses approximate network datasets built from TIGER street centerlines that are free for use.

Tutorial 9-1: Using buffers for proximity analysis

A buffer is a polygon surrounding map features of a feature class. As the analyst, you will specify a buffer's radius, and the **Buffer tool** will sweep the radius around each feature, remaining perpendicular to features, to create buffers. For points, buffers are circles; for lines, they are areas or rectangles with rounded endpoints; and for polygons, they are enlarged polygons with rounded vertices.

  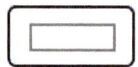

Generally, you use buffers to find what's near the features being buffered. An example of spatial analysis with buffers is so-called food deserts, which are areas that are more than a mile from the nearest grocery store in a city. Sometimes, buffers are exactly the right tool. One such example is the drug-free zones around schools for which federal and state laws prescribe a buffer radius, generally 1,000 feet, for school properties. The example in tutorials 9-1 and 9-2 is access by youths to public swimming pools in Pittsburgh. The closer that someone is to a pool, the more likely that person is to use it.

Pittsburgh has an irregular street network because of its rivers and hilly terrain, so even though some youths appear to be close to a pool on a map, they may have no direct route to the pool. In this case, you will need a network model that uses travel distance on a street network dataset. Buffers estimated with street networks are called service areas, and you will work with them in tutorial 9-3 to analyze public swimming pools in Pittsburgh.

Run the Pairwise Buffer tool

In this tutorial, you will buffer Pittsburgh's 32 public pools to estimate the number of youths ages 5 to 17 that live close, within a half mile, of the nearest pool. City block centroids, on the map, are symbolized with the number of youths.

1. Open **Tutorial9-1.aprx** from **Chapter9\Tutorials** and save it as **Tutorial 9-1YourName.aprx**.

2. Search for and open the **Pairwise Buffer** tool. Apply these settings and run the tool:
   - **Input Features**: Pools
   - **Output Feature Class**: **Pools_Buffer**
   - **Distance**: **0.5** US Survey Miles
   - **Dissolve Type**: Dissolve all output features into a single feature

This option dissolves interior lines of overlapping buffers, merging them into a single buffer.

Select block centroids within buffers and sum the number of youths

1. On the **Map** tab, in the **Selection** group layer, click **Select By Location** and apply these settings:
   - **Input Features**: PittsburghBlockCentroids
   - **Selecting Features**: Pools_Buffer

2. Click **OK**.

3. Open the **PittsburghBlockCentroids** table, right-click **AGE_5_17**, and click **Summarize**. Apply the following settings:
   - **Field**: AGE_5_17
   - **Statistic Type**: Sum
   - Clear the **Case** field

4. Click **OK**.

5. Open the **PittsburghBlockCe_Statistics** table.

Across the city, approximately 21,833 youths are within a half mile of a public pool. There are 48,903 youths in the city, and 44.6 percent are close to a pool.

6. Close the tables and clear the selection.

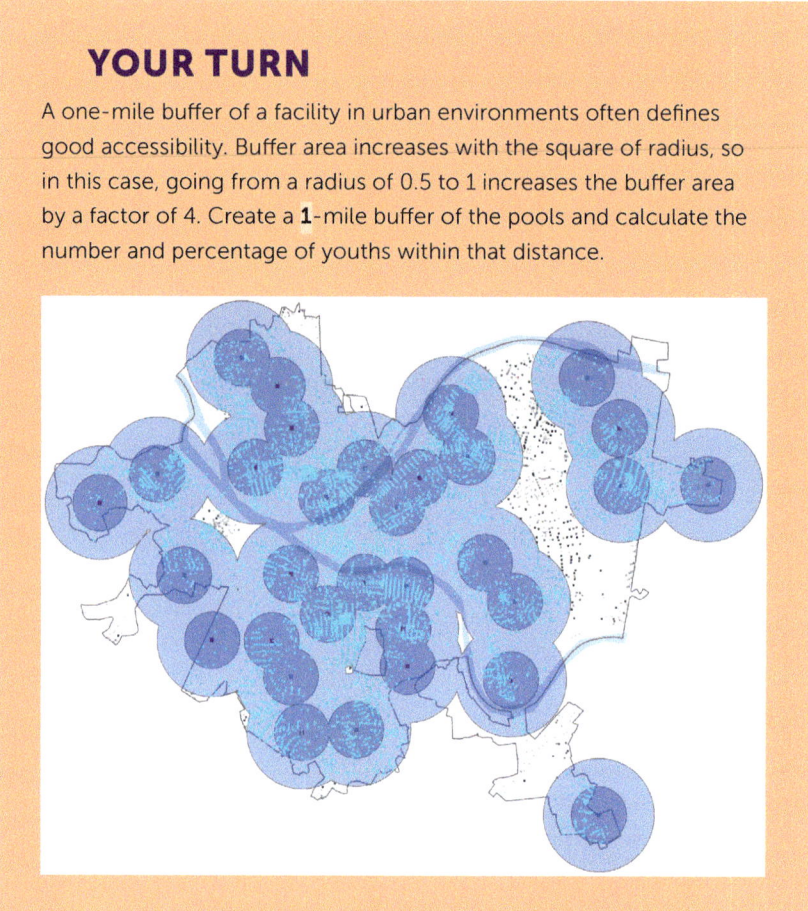

YOUR TURN

A one-mile buffer of a facility in urban environments often defines good accessibility. Buffer area increases with the square of radius, so in this case, going from a radius of 0.5 to 1 increases the buffer area by a factor of 4. Create a **1**-mile buffer of the pools and calculate the number and percentage of youths within that distance.

Tutorial 9-2: Using multiple-ring buffers

A multiple-ring buffer for a point looks like a bull's-eye target, with a center circle and rings extending out. You can configure the center circle and each ring to be separate polygons, thus allowing you to select other features within given distance ranges from the buffered feature.

During a budget crisis, Pittsburgh officials closed 16 of 32 public swimming pools across the city. You will estimate the number of youths ages 5 to 17 living at different distances from the nearest swimming pool for all 32 pools versus the 16 that were kept open. Youths living within a half mile of the nearest open pool are considered to have excellent access to pools, whereas youths living from a half to one mile from the nearest pool are considered to have good access (walkable in a half hour or less). Youths living farther than one mile from the nearest pool are considered to have fair to poor access (borrowing from the definition of food deserts). In tutorial 9-3, you will make more precise access estimates based on travel time across the street network of Pittsburgh from youth residences to the nearest pool.

Create multiple-ring buffers

1. Open **Tutorial9-2.aprx** from **Chapter9\Tutorials** and save it as **Tutorial 9-2YourName.aprx**.

   The map has all 32 public pools (symbolized by open and closed) and block centroids symbolized with youth population, ages 5 to 17. You will obtain the number of youths who had good access when only 16 pools were open.

2. Open **Select By Attributes** and apply these settings:
   - **Input Rows**: Pools
   - **Create the expression**: Where OPEN is equal to 1

3. Click **OK**.

4. Search for and open the **Multiple Ring Buffer** tool. Apply these settings and run the tool.
   - **Input Features**: Pools
   - **Distances**: **0.5**
   - Click **Add Another** to add an additional buffer ring
   - **Distances**: **1**
   - **Distance Unit**: Miles

   Tip: The default **Dissolve Option**, **Non-overlapping (rings)**, is needed for analysis and display. It merges overlapping buffer rings.

5. Compared with the analysis of pool access in tutorial 9-1, with all pools open, this section shows the significant reduction in access with only half the pools open. Next, you'll obtain the corresponding statistics.

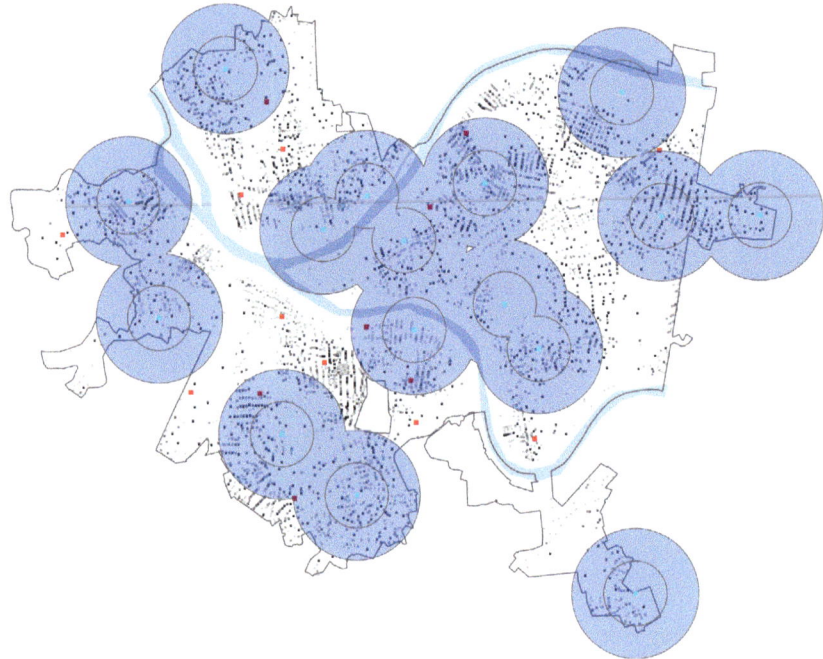

Use spatial overlay to get statistics by buffer area

You will run the **Spatial Join** tool, which will join all attributes of the multiple-ring buffer to block centroids and sums the number of youths ages 5 to 17 inside each ring.

1. Search for and open the **Spatial Join** tool. For **Target Features**, select **Pools_MultipleRingBuffer**. For **Join Features**, select **PittsburghBlockCentroids**.

2. Expand **Fields** and click **Edit**. In the **Fields** list, click **AGE_5_17**. In the **Table** list, click **PittsburghBlockCentroids (1)**. Under **Actions and Source Fields**, open the drop-down menu and select **Sum**. Click **OK**.

3. Open the resulting layer's attribute table to see in the **AGE_5_17** attribute how many youths had excellent access (0.5 distance row) and how many had good access (1 distance row).

   This calculates the percentage of youths with excellent (0- to 0.5-mile buffer) and good (0.5- to 1-mile buffer) access with the limited number of pools open. Recall that the total number of Pittsburgh youths is 48,903. Is there much difference in access for youths in the good (0.5- to 1-mile) buffer when all pools are open versus when only half are open? Why do you think that is?

> **YOUR TURN**
>
> Suppose that policymakers decide to add one more buffer that they propose to call fair access. Run the **Multiple Ring Buffer** tool with three buffers, adding an additional buffer at **1.5** miles. Calculate the percentage of youths in that buffer. Should the city open additional pools to provide these youths with easier access?

Tutorial 9-3: Creating multiple-ring service areas for calibrating a gravity model

Service areas are like buffers but are based on travel over a network, usually a street network dataset. If a point (say, for a retail store) has a five-minute service area constructed using **Network Analyst** tools in ArcGIS Pro, anyone residing in the service area has, at most, a five-minute street trip to the store using the shortest route.

If you have permission from your instructor to use an ArcGIS Online service that consumes purchased credits or otherwise have access to your own ArcGIS Online credits, you could use an ArcGIS Online network service. Such a service would be more accurate than the free, TIGER-based network dataset (**PittsburghStreets_ND**) that you will use in this chapter.

In this tutorial, you will use service areas to estimate a gravity model of geography, which assumes that the farther apart two features are, the less attraction between them. The falloff in attraction with distance is often nonlinear and rapid, as in Newton's gravity model for physical objects, where the denominator of attraction is distance squared.

The application of this tutorial is a continuation of the pool case study, based on a random sample of youths owning pool tags (which allow admission to any Pittsburgh public pool). The sample of individual youths with pool tags allows you to estimate a gravity model for pool use demand.

This tutorial uses the following multiple-step workflow:
1. Select open pools from the feature layer of all pools (so that the service area of the next step is applied to only open pools).
2. Create service areas for a given set of travel times (1, 2, 3, 5, and 7 minutes) to produce nonoverlapping service-area rings. The result is the center area and four rings (0 to 1 minute, 1 to 2 minutes, and so on). Three minutes' travel time corresponds roughly to a one-mile-radius buffer. For example, a car averaging 20 miles per hour in a city can go one mile in three minutes.

3. Count the number of pool tags in each service area. The count is accomplished using a spatial join of the service areas and pool tag points. When scaling up the sample by multiplying by 11.3, the result is an estimate of the number of total youths owning pool tags in each service-area ring (the sample size fraction of the total number of youths with pool tags was 1/11.3).
4. Sum the population of youths ages 5 to 17 in each service area using a spatial join of the service areas and block centroids. The result is an estimate of the population of youths having pool tags in each service ring.
5. Calculate the average use rate for each service area as a percentage: use rate = 100 × 11.3 × (number of youths with pool tags in sample) / (total population of youths).
6. Plot the estimated use rate of each service area versus the average travel time in each service area in minutes from the nearest pool (0.5, 1.5, 2.5, 4.0, 6.0). Use rate is expected to decline rapidly with travel time.
7. In tutorial 9-4, you will use your estimated gravity model to find the 16 best pools out of 32 to remain open in terms of maximizing youth pool attendance across Pittsburgh.

Create multiple-ring service-area polygons

1. Open **Tutorial9-3.aprx** from **Chapter9\Tutorials** and save it as **Tutorial 9-3YourName.aprx**.

   The map has all 32 public pools (both open and closed) and pool tags. The map also has the network dataset for Pittsburgh, **PittsburghStreets_ND**, built from TIGER street centerlines. The network dataset is calibrated for drive time in a vehicle.

   You will select the open pools, which of course were the only ones to have pool tags in the study period of pool closures.

2. Turn off **Pooltags**.

3. Open **Select By Attributes** and apply these settings:
   - **Input Rows**: Pools
   - **Expression**: Where OPEN is equal to 1

4. Click **OK**.

5. On the **Analysis** tab, in the **Workflows** group, click **Network Analysis** > **Service Area**.

ArcGIS Pro builds the **Service Area** group layer in the **Contents** pane and adds the **Network Analyst** service area context tab to the ribbon.

6. Click the **Service Area Layer** tab. In the **Input Data** group, click **Import Facilities** and apply these settings:
   - **Input Locations**: Pools
   - **Append to Existing Locations** check box: checked
   - **Snap to Network** check box: checked

7. Click **OK**.

The open pools become facilities for analysis using the street network.

8. In the **Travel Settings** group, for **Direction**, select **Toward facilities**.

ArcGIS Pro has detected that the street network, **PittsburghStreets_ND**, is set up for travel time in minutes and suggests 5, 10, and 15 minutes for cutoffs defining travel areas. You will change these times next.

9. For **Cutoffs**, type **1**, **2**, **3**, **5**, **7**.

10. In the **Output Geometry** group, change the **Boundary Type** to **Dissolve** and keep **Standard Precision** and **Rings** selected.

By selecting **Dissolve**, your service areas will provide travel time to the nearest pool.

11. On the **Service Area Layer** tab, in the **Analysis** group, click **Run**.

If you encounter an error stating that the tool is not licensed, license the tool following the steps outlined in the software requirements and licensing section in the book introduction.

12. Symbolize the output polygons, using the third row in the **ArcGIS Colors** list, so that **1** is blue, **2** is blue-green, **3** is green, **5** is yellow, and **7** is orange.

13. Turn on labeling for **Pools** and move that layer above **Service Area**. Clear any selected features.

Suppose that a travel time of less than one minute is excellent, one to two minutes is very good, two to three minutes is good, three to five minutes is fair, and more than five minutes is poor. Then according to estimated street travel time in a car, areas that are considered fair or poor are clearly visible in yellow and orange with two interesting results. First, the service areas are different from circular pool buffers because Pittsburgh's streets have irregular patterns caused by the city's many hills and valleys. **Jack Stack** and **Highland** pools are good examples. Zoom in to their areas to see their streets. Second, you will see open pools close together, such as **Cowley** and **Sue Murray**, where perhaps one could be closed. Also, closed pools, such as **Fowler**, could be opened to provide better access. The analysis that you just performed was not available at the time of initial pool closings; also, city officials used criteria in addition to geographic access when closing pools, such as the condition of pools and historical attendance patterns.

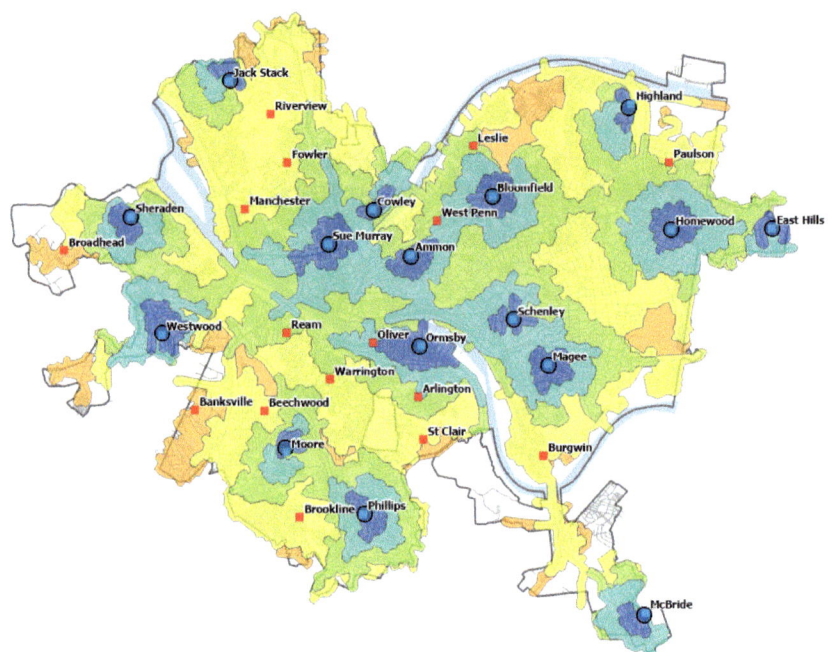

Next, you will count youths with pool tags and sum youth population, all by service-area polygons.

Spatially join service areas and pool tags

Each youth has a pool tag, so counting provides the number of potential pool users in each service area.

1. Turn on **Pooltags**.

2. Open the **Spatial Join** tool. For **Target Features**, select **Service Area\Polygons**. For **Join Features**, select **Pooltags**.

3. Expand **Fields** and click **Edit**. In the **Fields** list, click **FacilityID** and change **Actions and Source Fields** to **Count**. Click **OK**.

4. Click **Run**.

   This rule counts nonnull values of any attribute, so the field (**FieldID**), which is present for all records and is nonnull, is a good choice.

5. Open the attribute table for **Polygons_SpatialJoin**.

   In the **Join_Count** field, there are 374 youth tag holders in the zero- to one-minute service area, 512 in the one- to two-minute area, and so on. You must change the name of the **Join_Count** field so you can identify its purpose.

6. In the **Polygons_SpatialJoin** attribute table, click **Options** > **Fields View**, change the **Field Name** and **Alias** of Join_Count to **Join_Count_Pooltags**, save the changes, and close all tables.

   The purpose of the next join is to estimate the number of youths in each service area so that you can then estimate the fraction of youths that have pool tags in each service area.

7. Open the **Spatial Join** tool. For **Target Features**, select **Polygons_SpatialJoin**. For **Join Features**, select **PittsburghBlockCentroids**. For **Output Feature Class**, type **Polygons_Tags_Pop**.

8. Expand **Fields** and click **Edit**. In the **Fields** list, click **AGE_5_17** and change **Actions and Source Fields** to **Sum**. Click **OK**.

9. Click **Run**.

   For blocks with centroids within service areas, this rule sums corresponding youths ages 5 to 17.

   The tool creates a new feature class, **Polygons_Tags_Pop**, which is added to the **Contents** pane. The **AGE_5_17** attribute of this feature class is an estimate of the number of youths in each service area.

Calculate pool use statistics for service areas

Next are the final steps of the workflow—calculating and plotting use rate.

1. Open the **Polygons_Tags_Pop** attribute table and open the **Fields** view.

2. Create two fields: **AverageTime** and **UseRate**, both with the **Float** data type.

   AverageTime stores the midpoint of each service area ring—for example, 0.5 for the 0-1 ring.

3. Move the **Name** column directly before the **AverageTime** field.

4. Save and close the **Fields** view.

   In the **Polygons_Tags_Pop** table, type these values for **AverageTime**, referring to the **Name** column to align with the correct travel time ranges: **6**, **4**, **2.5**, **1.5**, and **0.5**. On the **Edit** tab, save your edits.

Name	AverageTime
5 - 7	6
3 - 5	4
2 - 3	2.5
1 - 2	1.5
0 - 1	0.5

   The final task is to compute estimated use rates for each service-area ring, using the following expression:

   ```
   100 × 11.3 × !Join_Count_Pooltags!/!Age_5_17!
   ```

   In this expression, 100 makes the result a percentage, 11.3 is the scale-up factor from the random sample to the population level, **!Join_Count_Pooltags!** is the sample number of youth tag holders in each service-area ring, and **Age_5_17** has the total (sum) of youths in each ring.

5. Right-click **UseRate** and click **Calculate Field**. After **UseRate** =, enter the following expression:

   ```
   100 * 11.3 * !Join_Count_Pooltags! / !AGE_5_17!
   ```

6. Click **OK**.

As expected, use rate declines quickly with average time from the nearest pool: 91.3 percent of youths within a mile of the nearest pool have pool tags, but the rate drops quickly, all the way down to 25.8 percent in the five- to seven-minute ring.

7. Close the attribute table.

Make a scatterplot

In ArcGIS Pro, you will make a scatterplot of **Use Rate versus Average Time** to visualize the estimated gravity model data points.

1. In the **Contents** pane, click **Polygons_Tags_Pop**.

2. On the feature layer's **Data** tab, in the **Visualize** group, click **Create Chart** > **Scatterplot**.

3. In the **Chart Properties** pane, for **X-axis Number**, select **AverageTime**; for **Y-axis Number**, select **UseRate**.

   The gravity model curve falls rapidly in a nonlinear way as expected. You can change the title and axis labels of the chart using the **General** tab of the **Chart Properties** pane. Automatically included in the chart is a straight-line regression model, which is not an appropriate form for the evidently nonlinear relationship.

4. Save your project.

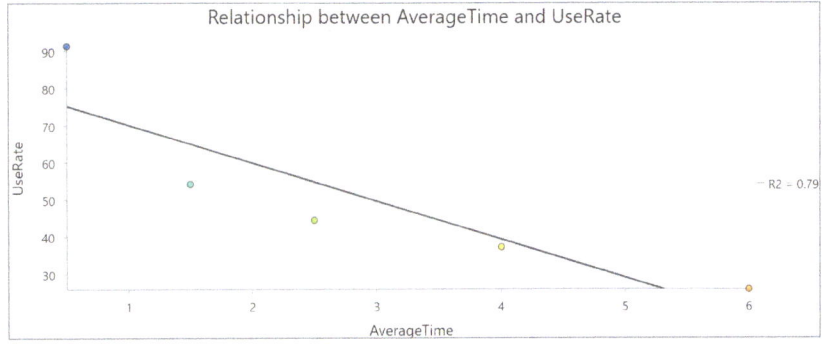

Fit a curve to the gravity model data points

Next, you need to fit a curve to the five points of the scatterplot to yield a gravity model function for use in the location-allocation model. The location-allocation model finds the best facility locations from a set of candidate locations, using a gravity model for use rates, which are fractions of target populations that will use facilities.

Available in the location-allocation model are three functional forms for a gravity model: linear, exponential, and power. The power form does not work well for cases with short travel times, such as for the swimming pools (a few minutes), so it is not discussed further here. The linear form is a poor fit for the available data. The exponential form is the most applicable for the swimming pool case because it declines rapidly as travel time increases. The Microsoft Excel worksheet, **Exponential.xlsx**, available in **Chapter9\Data**, provides a tool for fitting the exponential model to the points in the scatterplot.

1. Open **Exponential.xlsx** from **Chapter9\Data**.

   Beta, which currently has the value 0.25, is the single parameter that you must choose (optimize) in the worksheet and provide in the location-allocation model. Impedances are average travel times in minutes. The **Gravity Model Use Rate** column has the estimates between 9 and 3 for its multiring service areas, expressed here as fractions. **Cost** is an exponential function, $e^{\beta T}$, used in the network optimization model to represent system impedance (resistance to using a facility), and the model uses Pro Use Rate, $e^{-\beta T}$, as the gravity model. Your task is to vary beta until you get a good fit for the **Pro Use Rate** to the **Gravity Model Use Rate**. **Average Absolute Error** is a guide for choosing beta; you must minimize it by trial and error.

2. Type the following values for **Beta**, one at a time, in the worksheet: **0.10**, **0.15**, **0.20**, **0.25**, **0.30**, and **0.35**, and note resulting average absolute error values.

   The results are saved to the worksheet in rows 13 through 19. Beta = 0.25 is the best fit from the values tried. If you are comfortable with Microsoft Excel, you can modify this worksheet for other case studies or projects by adding (copying) rows and changing impedances.

3. Save and close the worksheet. Save your project.

Tutorial 9-4: Using Network Analyst to locate facilities

Suppose that you are an analyst for an organization that owns several facilities in a city, and you are asked to find the best locations for new facilities. The classic problem of this kind is to locate facilities for a chain of retail stores in an urban area, but other examples are federally qualified health centers (FQHCs), as you learned in chapter 1, and the public swimming pools you worked with in this chapter. In the most general case, your organization has existing facilities that will remain open, a set of competitor facilities, and a set of potential new locations from which you want to determine a best subset of a specified size. Another case: A subset of existing facilities must be closed, as with the Pittsburgh swimming pools, for which you want to determine the appropriate 16 of 32 facilities to close. Yet another case: There are no existing facilities, and you want to locate one new facility.

In ArcGIS Pro, the location-allocation model in the **Network Analyst** collection of models handles this sort of facility location problem. Inputs are a network dataset, locations of facility types (existing, competitors, and new potential sites), demand points, and a gravity model. Demand is represented by polygon centroids—blocks, block groups, tracts, zip codes, and so on—for which you have data on the target population, generally from the US Census Bureau, such as youth population. Resistance to flow in the network, known as impedance, is represented by a gravity model and can be distance or time traveled along shortest paths to facilities. A network optimization model selects a specified number of new facility locations that maximize attendance.

In this tutorial, you will run a model to choose the best 16 of 32 swimming pools to keep open, using geographic access (distance from the nearest pool) as the criterion.

Find the 16 of 32 pools that maximize attendance

1. Open **Tutorial9-4.aprx** from **Chapter9\Tutorials** and save it as **Tutorial 9-4YourName.aprx**.

   This map has the 32 pools, block centroids with youth population, and the **PittsburghStreets_ND** network dataset. Also included is the **PittsburghStreets** street centerline layer symbolized using a tan color for visualization. The **PittsburghBlockCentroids** table has an attribute, **Cutoff_Minutes**, which has value **3**, meaning that any block centroid farther than three minutes' travel time to the nearest pool under consideration will not be served—thus, none of its youth will go to a pool. Recall from tutorial 9-3 that a travel time by car of less than a minute is excellent, one to two minutes is very good, and two to three minutes is good. So, the analysis of this tutorial will be for good to excellent pool access.

2. On the **Analysis** tab, in the **Workflows** group, click **Network Analysis** > **Location-Allocation**.

   ArcGIS Pro creates the **Location-Allocation** group layer in the **Contents** pane.

3. Click the **Location-Allocation Layer** tab.

   The horizontal toolbar for this model opens. If this toolbar ever closes and you want to reopen it, select the **Location-Allocation** group layer in the **Contents** pane.

4. In the **Input Data** group, click **Import Facilities** and apply these settings:
   - **Input Locations**: Pools
   - **Property**: FacilityType
   - **Default Value**: Candidate
   - **Append to Existing Locations** check box: checked

   Tip: You can run this tool more than once to load different kinds of facilities.

5. Click **OK**.

   All 32 pools become candidates for selection to keep open.

6. In the **Input Data** group, click **Import Demand Points** and apply these settings:
   - **Input Locations**: PittsburghBlockCentroids
   - **Property**: Weight
   - **Field Name**: AGE_5_17

7. Click **OK**.

   Wait for the 7,493 demand points to appear.

8. In the **Travel Settings** group, for **Direction**, select **Towards facilities**, and for **Facilities**, type **16**.

9. In the **Problem Type** group, for **f(cost, β)**, select **Exponential**. For β, type **0.25** and press **Tab**.

10. Run the model.

    Wait for the model to run.

11. In the **Contents** pane, in the **Location-Allocation** group layer, turn off **Demand Points** and resymbolize **Lines** to have a **Line width** of **0.5** and a **Color** of your choice.

    The 16 optimal pools that maximize attendance have a star on their point symbols. The lines connecting pools and served demand points constitute a spider map. This sort of map is used for visualization—these lines show the demand relationship between pools and block centroids.

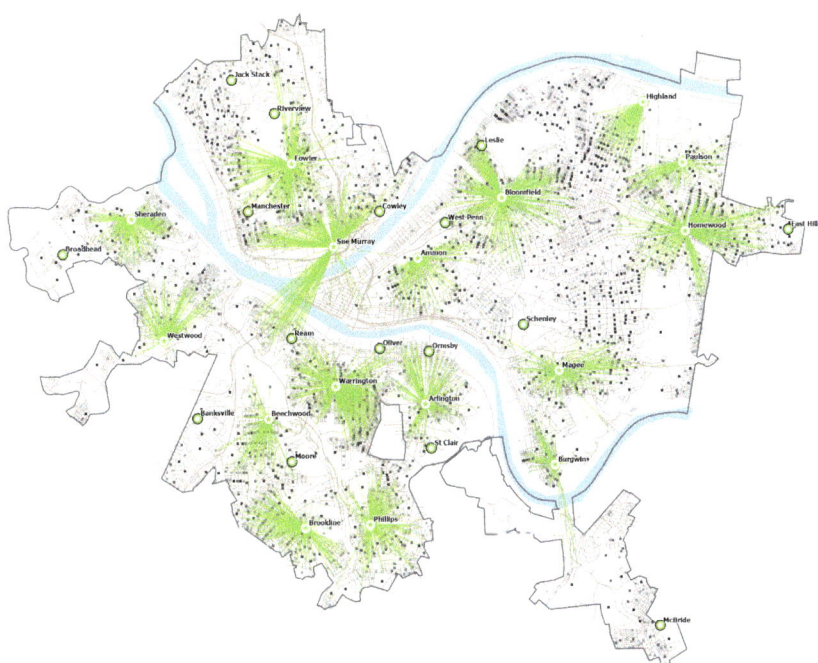

Analyze the optimal solution

1. Open the **Facilities** attribute table and sort **DemandWeight** in descending order.

    DemandWeight has the number of youths allocated by the location-allocation model to each pool. However, only about 50 percent of Pittsburgh's youths had pool tags in the study year, so a better estimate of the number of users for each pool is to divide the **DemandWeight** values by 2. Then, for example, you would expect 1,896/2 = 948 users of the Sue Murray pool.

2. Open the **Pools** attribute table.

   MAXLOAD is the capacity of a pool. The **Sue Murray** pool has a capacity of 700, so an expected demand of 948 is overcapacity. However, not every potential user of the pool will show up at once, so this solution may be feasible.

3. In the **Facilities** attribute table, use the **Summarize** tool to sum **DemandWeight**.

   The total is 34,154, which is the total number of youths with good access to the nearest pool. Dividing by 2, the estimate is that 17,077 youths would use the 16 optimal pools under existing conditions. Perhaps more youths with good access would use pools in the future.

4. Close all tables and save your project.

> **YOUR TURN**
>
> Compare the optimal solution to the existing solution that officials chose. First, select open pools with the clause **Where OPEN is equal to 1**. In the **Contents** pane, turn off the **Location-Allocation** group layer. Create a new **Location-Allocation** model and run **Import Facilities** using **Pools** for **Input Locations**, **FacilityType** for **Property**, and **Required** for **Default Value**. Then, run **Import Demand Points** using the same parameters earlier in the tutorial. Run the **Location-Allocation** model using the same configurations earlier in the tutorial. Open the table for **Facilities** and sum **DemandWeight**. You should find that 26,536 youths had good access to the open pools, compared with 34,154 in the optimal solution.
>
> If you have time, make one more model run, this time using all 32 pools. You will find that the model estimates 39,846 youths with good access to all 32 pools, compared with 34,154 for the optimal solution with 16 open pools. You can conclude that the optimal solutio n is quite good, but the existing one falls short. In the year following the closure of pools, officials tasked one of the authors of this book to perform location analysis and adjust the set of open pools to improve citywide access for youths.

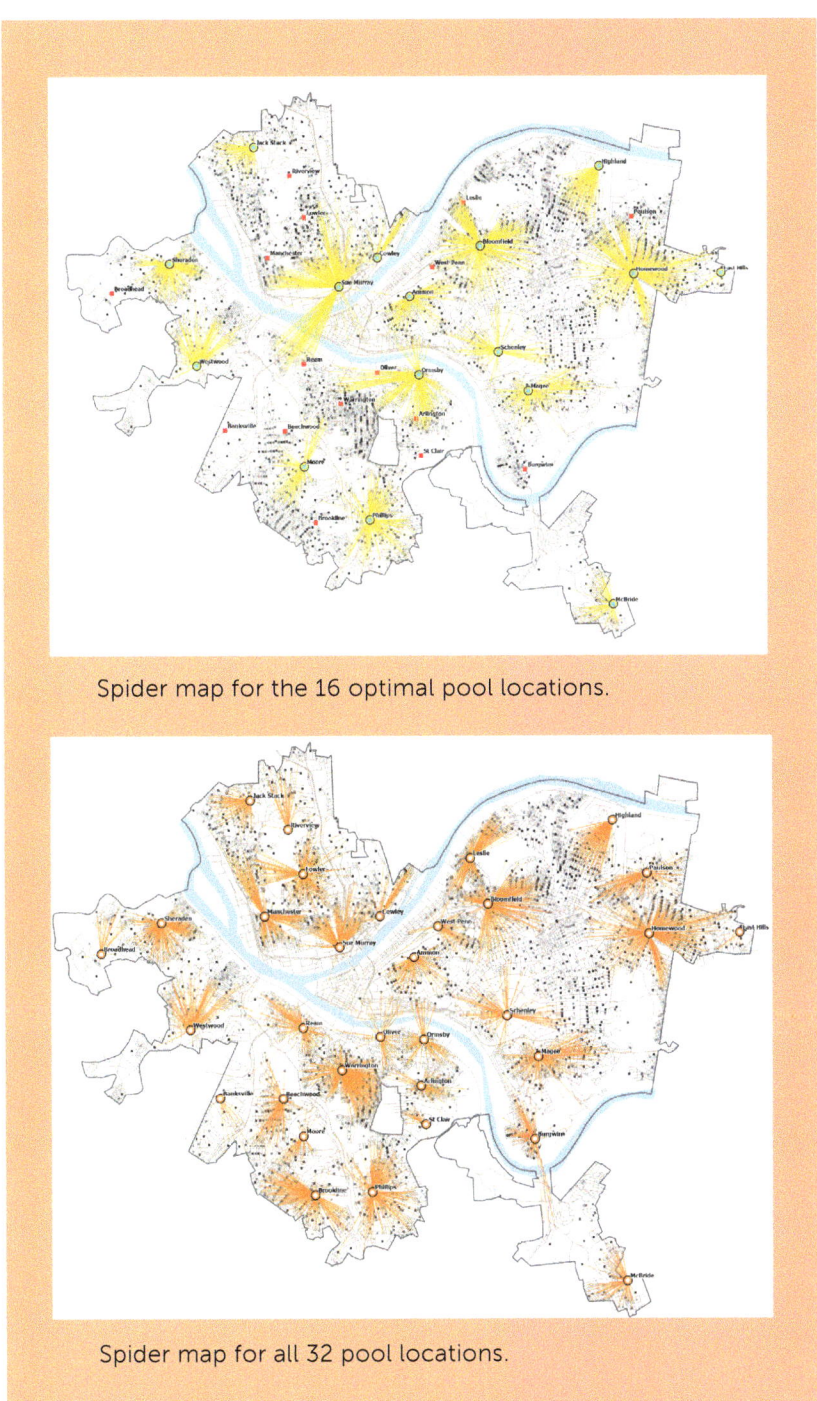

Spider map for the 16 optimal pool locations.

Spider map for all 32 pool locations.

Tutorial 9-5: Performing data cluster analysis

The goal of data mining is exploration—to find hidden structure in large and complex datasets that has some interest or value. Data clustering, a branch of data mining, finds clusters of data points that are close to each other but distant from points of other clusters. If your data points were 2D and graphed as a scatterplot, you could easily draw boundaries around points and call them clusters. The problem is that when the points lie in 3D space, you cannot see the clusters anymore. In this tutorial, you will cluster crimes using three attributes or dimensions, including severity of crimes, plus the age and gender of arrested persons.

A limitation of clustering is that there is no way of knowing true clusters in real data to compare with what an algorithm determines are clusters. You take what clustering methods provide, and if you get information or ideas from cluster patterns in your data, you can confirm them or determine their predictability using additional models or other methods. Clustering is purely an exploratory method. There are no right or wrong clusters—only more or less informative ones.

This tutorial uses k-means clustering, a simple method available in the **Multivariate Clustering** tool. K-means clustering partitions a dataset with n observations and p variables into $k < n$ clusters. In the tutorial, you will use a dataset with $n = 303$ observations, $p = 3$ variables for clustering, and $k = 5$ clusters. K-means is a heuristic algorithm, as are all clustering methods: It is a set of repeated steps that produces good, if not optimal, results. For this tutorial's data, the algorithm starts with a procedure that selects five 3D observations, called seeds, as initial centroids for clusters. Each observation is assigned to its nearest centroid, based on Euclidean (straight line) distance in the 3D cluster variable space. Next, a new centroid is calculated for each cluster, and all observations are reassigned to the nearest centroid. These steps are repeated until the cluster centroids do not move appreciably. K-means clustering can be thought of as a common-sense method.

K-means assumes that all attributes are equally important for clustering because it uses distance between numeric points as its basis. To meet this assumption, it is important that you scale all input attributes to similar magnitudes and ranges. Generally, you can use standardized variables (for each variable, subtract its mean and divide by its standard deviation) to accomplish scale parity, but other ways of rescaling are acceptable, too. It is the range, or relative distances between observations, that is important in k-means clustering. The data used in this tutorial includes numeric (interval or ratio), ordinal, and nominal class data for classification, whereas, strictly speaking, k-means clustering is intended for numeric data because of its use of distance in cluster variable space. Nevertheless, with rescaling, it is possible to get informative clusters when including nonnumeric data.

The data used in this tutorial is for serious violent crimes from a summer in Pittsburgh, with the data mapped as points. The crimes are ranked by seriousness using FBI hierarchy numbers: 1 = murder, 2 = rape, 3 = robbery, and 4 = aggravated assault. The nature of crimes should be important for their clustering. So, the first assumption to make is that the distance between crime types, such as 3 between 1 for murder and 4 for aggravated assault, is meaningful for clustering purposes. The criminal justice system agrees on the order, and for clustering, you can accept the distances or change them using your judgment. You will leave them as listed here.

Next, consider the single numeric attribute that is available: age of arrested person. Crime is generally committed by young adults, tapering off with older ages. For the serious violent crimes studied here, age varies between 13 and 65 years (range of 52), with a mean of 39. Together with crime seriousness, age would dominate clustering because of its greater range. The remedy is to standardize age so that it varies between −2.3 to 2.7, with a range of 5, whereas crime seriousness has a range of 3. This would make both attributes similar in determining clusters.

Finally, there is a nominal attribute, gender (male or female). You can encode this attribute as a binary attribute: 0 = male, 1 = female. As a binary indicator variable, gender has a mean, which is the fraction of arrestees who are female. This variable as encoded would have perhaps a lesser role than the previous two, but not by much. If you wanted to increase the importance of the binary variables for clustering, you could encode it as (0, 2) or (0, 4) indicators. You will leave it as a (0, 1) variable, which makes interpreting clustering results easier.

One last point is that you must choose the number of clusters instead of expecting k-means clustering to find an optimal number for you. This is the case for most clustering methods. For the crime data, experimentation with three to six clusters resulted in five clusters being the most informative, so you will use five clusters.

In summary, each observation is a 3D vector (crime, standardized age, gender). The clusters found by k-means exist in the 3D space in which the observations lie. Each cluster is characterized by its centroid, with the corresponding means of each cluster variable.

Perform a cluster analysis

1. Open **Tutorial9-5.aprx** from **Chapter9\Tutorials** and save it as **Tutorial 9-5YourName.aprx**.

   The map shows the spatial distribution of serious violent crimes by crime type within police zones. Also shown are the poverty areas. Generally, there is a positive correlation between poverty and violent crime. Police do not record any measures of poverty for arrested persons, such as annual income, so you cannot readily include poverty as a clustering variable. The map adds poverty as an additional consideration for interpreting clustering results.

2. Open the attribute table for **Serious Violent Crimes**, scroll to the last column (**Seed**), and sort the data in descending order.

The five records with **Seed** value of **1**, found in a previous run of the multivariate clustering algorithm, are used as initial cluster centroids. The **Hierarchy** attribute is the FBI code for crime types and one of the three cluster attributes. **ArrAge** is the age of the arrested person and is the second cluster variable. Multivariate clustering automatically standardizes numeric variables, so **ArrAgeStnd**, which is **ArrAge** standardized, is not needed. **ArrGender** is the remaining cluster variable. There is a good deal of variation in the cluster variables of the seed records, which is desirable for clustering seeds.

3. Close the table.

   The k-means algorithm is in the **Multivariate Clustering** tool, which you will run next.

4. Search for and open the **Multivariate Clustering** tool, apply these settings, and run the tool.
   - **Input Features**: Serious Violent Crimes
   - **Output Features**: SeriousViolentCrimes_Clusters
   - **Analysis Fields**: Hierarchy, ArrAge, ArrGender
   - **Clustering Method**: K-means
   - **Initialization Method**: User-defined seed locations
   - **Initialization Field**: Seed

   Tip: If you were running this tool for your own data or in an assignment, you would use the optimized seed locations method instead of the user-defined seed locations for initialization method.

   When the tool has finished, the **SeriousViolentCrimes_Clusters** layer, which has the five identified clusters, is added to the **Contents** pane.

5. Turn off the **Serious Violent Crimes** layer.

> **YOUR TURN**
>
> As an aid in analyzing the resulting clusters, run the **Summary Statistics** tool and calculate the means for clustered variables by cluster. Find the mean for the statistics fields: **Hierarchy**, **ArrAge**, and **ArrGender**. For **Case Fields**, use **Cluster ID**.
>
> | Cluster ID | FREQUENCY | MEAN_Hierarchy | MEAN_ArrAge | MEAN_ArrGender |
> |---|---|---|---|---|
> | 1 | 75 | 2.64 | 22.706667 | 0 |
> | 2 | 2 | 1 | 51 | 1 |
> | 3 | 63 | 3.603175 | 45.984127 | 0 |
> | 4 | 70 | 4 | 24.428571 | 0 |
> | 5 | 55 | 3.818182 | 29.418182 | 1 |

Interpret clusters

Table 9-1 has the rows and values of the **Summary Statistics** table with labels added for ranges of cluster variable means.

Table 9-1. Cluster IDs with frequency, crime, age, and gender

ClusterID	Frequency	Crime	Age	Gender
1 Young males moderate	75	2.6 moderate	23 young	0 male
2 Middle-aged females highest	2	1.0 highest	51 middle age	1 female
3 Middle-aged males lowest	63	3.6 low	46 middle age	0 male
4 Young males lowest	70	4.0 lowest	24 young	0 male
5 Young females lowest	55	3.8 lowest	29 young	1 female

These results have moderately interesting patterns and one anomalous group, 2. With a group size of only two crimes for group 2, you cannot rely on the result that women in their early 50s or thereabouts are especially dangerous murderers. That is not a pattern likely to be repeated. Group 1 is young males committing a range of serious violent crimes. Group 3 is middle-aged criminals committing crimes toward the lesser end of serious violent crimes, mostly aggravated assaults (FBI hierarchy 4). Group 4, young males, is committing aggravated assaults. Finally, group 5, young females, mostly commits aggravated assaults. You will examine whether there are any spatial patterns for these groups.

1. Open the **Symbology** pane for **SeriousViolentCrimes_Clusters**, and in the **Label** column, relabel the groups as follows:
   - 1 to **Young males middle**
   - 2 to **Middle-aged females highest**
   - 3 to **Middle-aged males lowest**
   - 4 to **Young males lowest**
   - 5 to **Young females lowest**

2. Change the point symbols for the three young groups to **Square 1**. Change **Color** to their original colors, so that **Young males middle** is blue, **Young males lowest** is yellow, and **Young females lowest** is purple, and change the **Size** to **5**.

   The cluster results that were judged moderately interesting earlier get more interesting when mapped. The serious violent crimes in Pittsburgh's central business district (indicated by the cluster of green dots in the center of the map near the river junction) are predominantly by middle-aged criminals. Crime patterns in central business districts of cities are often unique in part because most persons in those districts travel to them from their residences over some distances. Youths tend to commit crimes near where they live, whereas older criminals, who have greater mobility, travel to the central business district to commit crimes. Most crimes in poverty areas are committed by young persons, and the distance-to-crime theory of criminology states that these persons tend to commit crimes near where they live, within a mile or so. Somewhat surprising is the high percentage—21 percent—of the serious violent crimes in the data being aggravated assaults committed by females. Those crimes are scattered across Pittsburgh except the southwest police zone, which shows only one such crime. Of interest in that zone is the group of serious violent crimes committed by youths in an area not considered to be a poverty area (indicated by the cluster of blue and orange squares at the bottom of the map).

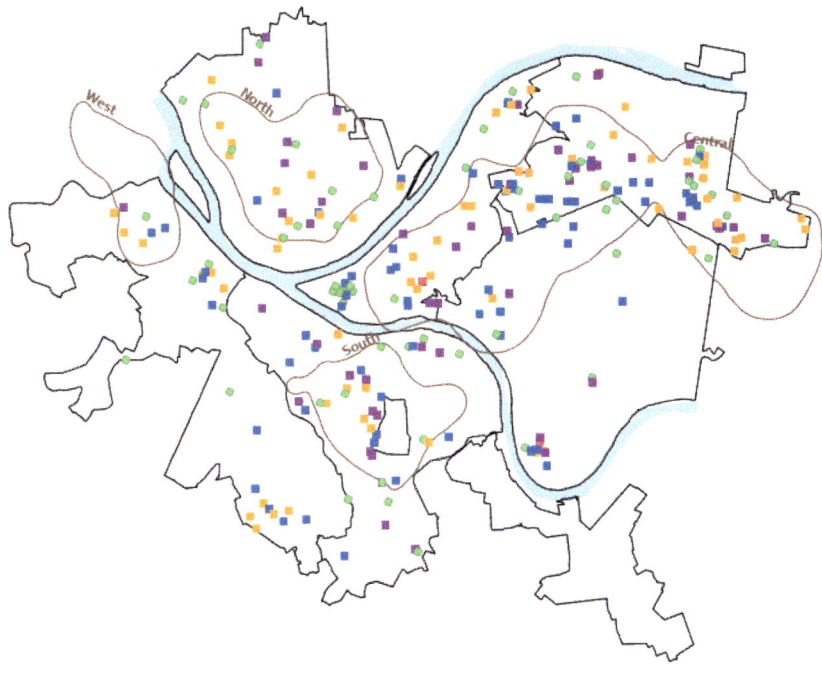

3. Save your project.

Assignments

This chapter has assignments to complete that you can download with data from ArcGIS Online at links.esri.com/GISTforPro3.4Assignments.

CHAPTER 10

Raster GIS

LEARNING GOALS

- Extract and symbolize raster maps.
- Create hillshade maps.
- Smooth point data with kernel density smoothing.
- Build a raster-based risk index.
- Build a model for automatically creating risk indexes.

Introduction

So far, this book has been devoted to vector feature classes (points, lines, and polygons), except for displaying an occasional raster layer. Vector feature classes are for discrete features (such as streetlight locations, streets, and city boundaries). Raster layers are for continuous features (for example, satellite images of Earth, topography, and precipitation). You can also use raster layers to display an attribute (for example, population) for large numbers of vector features (for example, city blocks, US counties). In such cases, you have so many vector features that choropleth and other maps are too small to render them clearly. As a remedy, as you will see in this chapter, you can transform vector feature classes into raster datasets that you can visualize.

Raster dataset is the generic name for a cell-based map layer stored on a disk in a raster data format. Esri supports more than 70 raster dataset formats, including familiar image formats such as TIFF and JPEG and GIS-specific formats such as Esri Grid. You can import raster datasets into file geodatabases.

All raster datasets are arrays of cells (or pixels)—each with a value and location and rendered with a color when mapped. As with any other digital image, the pixels are so small at intended viewing scales that they are not individually distinguishable. The coordinates for a raster dataset are the same kind used for vector maps (see chapter 5).

All raster datasets have at least one band of values. A band is comparable to an attribute for vector layers and stores the values of a single attribute in an array. The values can be positive or negative integers or floating-point numbers. You can use integer values for categories (codes), which must then have a layer file with descriptions and colors (for example, 1 = Agriculture, brown; 2 = Forest, green). Raster dataset values can also be floating-point numbers representing magnitudes (for example, temperature or slope of terrain).

Color capture and representation in raster datasets is an important topic. Color in the visible range is captured by satellites in three bands (red, blue, and green) that mix to produce any color. Color in many raster datasets, however, is often represented in one band using the colormap function, in which each color is given a code (integer value). Color depth is the number of bits (on/off switches in data storage) for code length used to represent colors. True color uses 24 bits per pixel and can represent more than 16 million colors (the human eye can distinguish about 10 million colors).

The spatial resolution of a raster dataset is the length of one side of a square pixel. If a pixel is one meter on a side, it has one-meter spatial resolution (a high resolution for high-quality maps). The US Geological Survey (USGS) provides imagery for urban areas in the United States at this resolution or higher, with images that can be zoomed in to small parts of neighborhoods (for example, with driveways, swimming pools, and tennis courts being clearly shown when displayed). The Landsat 7 and 8 satellites that together circle Earth every eight days have a resolution of 30 meters for most of their bands, which is good for viewing areas as small as neighborhoods of a city but not individual houses.

File sizes for raster datasets may require large amounts of disk space for storage and take a long time to process and display on a computer screen. Consequently, raster GIS uses several mechanisms to reduce storage and processing time, including data compression and pyramids. Pyramids provide reduced-resolution overviews at different scales that take less time to display than the original layer. A mosaic dataset is a data catalog for storing, managing, viewing, and querying collections of raster datasets, often forming a continuous map when viewed. Although such a mosaic dataset is viewed as a single mosaicked image, you also have access to each dataset in the collection. A mosaic dataset can store raster datasets of the same area for different times, allowing the display and comparison of different time periods.

Tutorial 10-1: Processing raster datasets

The ArcGIS Pro project that you will open has single-band raster datasets for land use and elevation, downloaded from the USGS website. You will extract raster datasets for Pittsburgh from each original dataset, which has extents larger than Pittsburgh. Because raster datasets are rectangular, you will display layers using Pittsburgh's boundary as a mask: Pixels in Pittsburgh's rectangular extent but outside the city's boundary will still exist but be given no color, whereas those within Pittsburgh will have assigned colors. Finally, you will use the elevation layer to produce a hillshade layer, which is a shaded relief rendering of topography created using an artificial sun to add illumination and shadows for a 3D appearance.

Examine raster dataset properties

1. Open **Tutorial10-1** from the **Chapter10\Tutorials** folder and save it as **Tutorial10-1YourName.aprx**.

   National Elevation Dataset (**NED**) is elevation data. It will look like topography after you create a hillshade for it. **LandUse_Pgh** needs symbology for interpretation.

   Raster datasets have considerable metadata that you can read as properties.

2. In the **Contents** pane, right-click **NED** and click **Properties**. Click the **Source** tab.

   The dataset is in the Chapter10 file geodatabase, where the dataset was imported. The elevation units are in feet above sea level.

3. Expand **Raster Information**.

   The raster dataset has one band with 2,106 columns and 1,984 rows, for a total of 4,178,304 cells. The cells have roughly 90-foot (30-meter) resolution. The raster has pyramids for speedy display when zooming out.

4. Scroll down and expand **Statistics**.

   The average elevation in the raster dataset is 323.7 feet, with a minimum of 206.9 feet and a maximum of 443.2 feet above sea level.

5. Expand **Extent**.

   The four values provided are state plane coordinates in feet that specify the four corners of the rectangular extent of the raster dataset.

6. Expand **Spatial Reference**.

   Here, the coordinates are in state plane, which uses a **Lambert Conformal Conic** projection tuned for southern Pennsylvania.

7. Click **Cancel**.

> **YOUR TURN**
>
> Review the properties of **LandUse_Pgh.tif**. The raster format is TIFF, a common image format. Its resolution is 30 meters, with a single band and a color map (which is available in a separate layer file that you will use in a section that follows). It has a projection for the 48 contiguous US states (**Albers Equal Area**) that distorts direction (which explains why it's tilted).

Import a raster dataset into a file geodatabase

Next, you will import **LandUse_Pgh.tif** into the Chapter10 file geodatabase.

1. Search for and open the **Raster to Other Format** tool. Apply these settings and run the tool:
    - **Input Rasters**: Browse to **Chapter10\Data**, click **LandUse_Pgh.tif**, and click **OK**.
    - **Output Workspace**: Browse to **Chapter10\Tutorials**, click **Chapter10.gdb**, and click **OK**.
    - **Raster Format**: Esri Grid.

2. Remove **LandUse_Pgh.tif** from the **Contents** pane, add **LandUse_Pgh** from **Chapter10.gdb**, and move it below **NED** in the **Contents** pane.

   Nothing appears to change on the map, but the format of **LandUse_Pgh** is now **Esri Grid**, and it is located inside a geodatabase.

Set the geoprocessing environment for raster analysis

Environmental settings affect how geoprocessing is carried out by tools in the current project. You will set the cell size of raster datasets you create to 50 feet, and you will use Pittsburgh's boundary as the default mask.

1. On the **Analysis** tab, in the **Geoprocessing** group, click **Environments**.

2. In the **Environments** settings, under **Raster Analysis**, for **Cell Size**, type **50**. For Mask, browse to **Chapter10.gdb** and select **Pittsburgh**.

3. Click **OK**.

Extract land use using a mask

Next, you will use the **Extract by Mask** tool to extract a new layer called **LandUse_Pittsburgh** based on the existing layer, **LandUse_Pgh**. The resulting raster dataset will have the same extent as Pittsburgh and therefore be a much smaller file to store than the original.

The **Extract by Mask** tool requires you to specify the mask layer explicitly if you want to override the default mask you set in **Environments**.

1. Search for and open the **Extract by Mask** tool. Apply these settings and run the tool:
    - **Input raster**: LandUse_Pgh
    - **Input raster or feature mask data**: Pittsburgh
    - **Output raster**: LandUse_Pittsburgh

    ArcGIS Pro applies an arbitrary color scheme to the new layer.

2. Remove **LandUse_Pgh** from the **Contents** pane.

3. In the **Catalog** pane, expand **Databases** and **Chapter10.gdb**, right-click **LandUse_Pgh**, and click **Delete** > **Yes**.

4. Zoom to the **Pittsburgh** bookmark.

> **YOUR TURN**
>
> Use the **Extract by Mask** tool to extract **NED_Pittsburgh** from **NED**. In the **Extract by Mask** tool pane, click **Environments** and the **Select Coordinate System** button on the right of **Output Coordinate System**. Expand **Projected Coordinate System** > **State Plane** > **NAD 1983 (2011) (US Feet)** > **NAD 1983 StatePlane Pennsylvania South FIPS3702 (US Feet)** and click **OK**. On the **Parameters** tab of the tool pane, choose **NED** as the input raster, **Pittsburgh** as the mask, **NED_Pittsburgh** as the output raster, and run the tool. After creating **NED_Pittsburgh**, remove **NED** from the map and delete it from **Chapter10.gdb**.

Symbolize a raster dataset using a layer file

1. In the **Contents** pane, move **LandUse_Pittsburgh** above **NED_Pittsburgh**.

2. For **LandUse_Pittsburgh**, open the **Symbology** pane, click the **Options** button (three lines), and click **Import from layer file**.

3. Browse to **Chapter10\Data** and double-click **LandUse.lyr**.

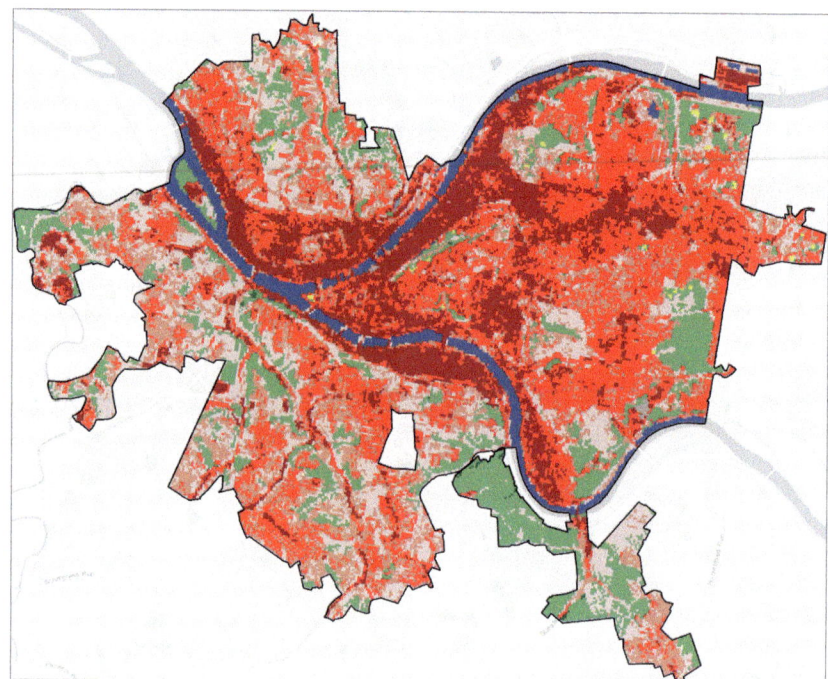

Create hillshade for elevation

Hillshade provides a way to visualize elevation. The **Hillshade** tool simulates illumination of the earth's elevation surface (here represented by the **NED_Pittsburgh** raster layer) using a hypothetical light source representing the sun. Two parameters of this function are the altitude (vertical angle) of the light source above the surface's horizon in degrees and its azimuth (angular direction). You will use the default values of the **Hillshade** tool for **Azimuth** and **Altitude**. The sun for your map will be in the west (315 degrees) at an elevation of 45 degrees above the horizon. The effect of applying hillshade to elevation data is striking because of its use of light and shadow. You can enhance the display of another raster layer, such as land use, by making it partially transparent and placing hillshade beneath it.

1. Search for and open the **Hillshade (3D Analyst Tools)** tool. Apply these settings and run the tool:
    - **Input raster**: NED_Pittsburgh
    - **Output Raster**: **Hillshade_Pittsburgh**

2. Turn off **NED_Pittsburgh**.

Symbolize hillshade

You can improve the default symbolization of **Hillshade_Pittsburgh**, as follows.

1. Open the **Symbology** pane for **Hillshade_Pittsburgh** and apply these settings:
   - **Primary Symbology**: Classify
   - **Method**: Standard Deviation
   - **Interval Size**: 1/4 standard deviation
   - **Color Scheme**: Black to White

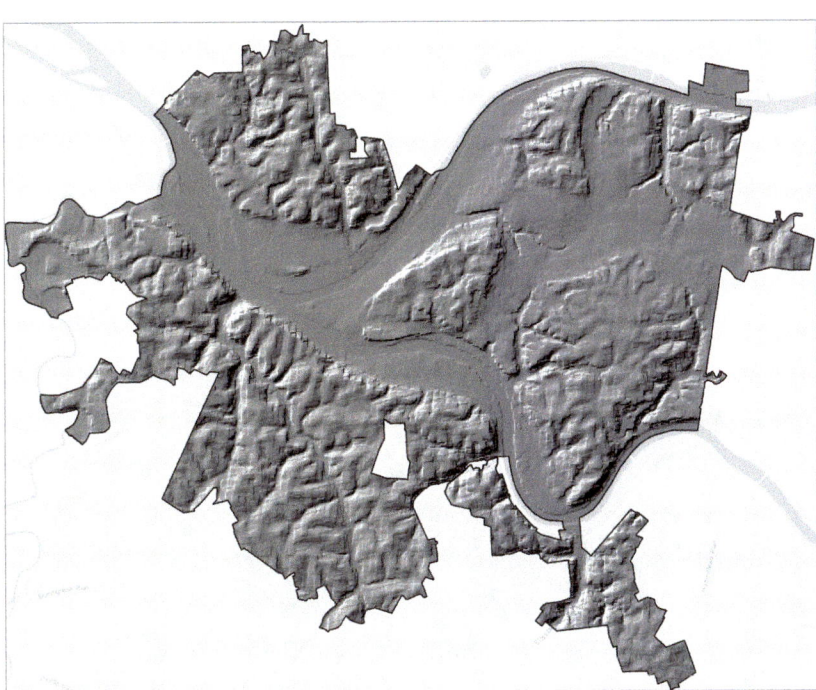

Use hillshade for shaded relief of land use

Next, you will make **LandUse_Pittsburgh** partially transparent and place **Hillshade_Pittsburgh** beneath it to give land use shaded relief.

1. Move **LandUse_Pittsburgh** above **Hillshade_Pittsburgh**.

2. Click the **LandUse_Pittsburgh** layer and click the **Raster Layer** contextual tab.

3. In the **Effects** group, for **Transparency**, type **33** and press **Enter**.

   Hillshade_Pittsburgh shows through the partially transparent **LandUse_Pittsburgh**, giving the land-use layer a rich, 3D-like appearance.

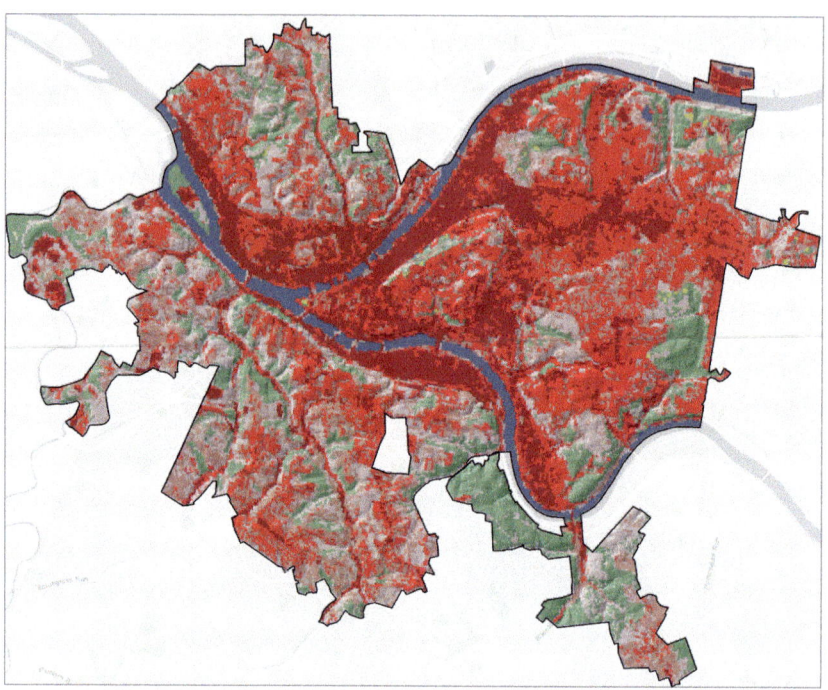

Create elevation contours

Another way to visualize elevation data is with elevation contours—lines of constant elevation commonly seen on topographic maps. For Pittsburgh, the minimum elevation is 215.2 feet, and the maximum is 414.4 feet, with about a 200-foot difference. If you specify 20-foot contours, starting at 220 feet, there will be about 10 contours. The output contours are vector line data and not polygon data.

1. Search for and open the **Contour** tool. Apply these settings and run the tool:
   - **Input raster**: NED_Pittsburgh
   - **Output feature class**: **Contour_Pittsburgh**
   - **Contour interval**: **20**
   - **Base contour**: **220**

2. Turn off **LandUse_Pittsburgh**, **Hillshade_Pittsburgh**, and **NED_Pittsburgh**.

3. Open the **Symbology** pane for **Contour_Pittsburgh** and change the current symbology to a medium-gray line with a **Line width** of **0.5**. Save your project.

   You can label contours with their contour elevation values when zoomed in, but you will not do that in this tutorial.

Tutorial 10-2: Making a kernel density (heat) map

Kernel density smoothing (KDS) is a widely used method in statistics for smoothing data spatially. The input is a vector point layer, often center points (centroids) of polygons for population data or point locations of individual demands for goods or services. KDS distributes the attribute of interest of each point continuously and spatially, turning it into a density (or heat) map. For population, the density is, for example, persons per square mile.

KDS accomplishes smoothing by placing a kernel, a bell-shaped surface with surface area 1, over each point. If there is population, N, at a point, the kernel is multiplied by N so that its total area is N. Then all kernels are summed to produce a smoothed surface, a raster dataset.

The key parameter of KDS is its search radius, which corresponds to the radius of the kernel's footprint. If the search radius is small, you will get highly peaked mountains for density. If you choose a large search radius, you will get gentle, rolling hills. If the chosen search radius is too small (for example, smaller than the radius of a circle that fits inside most polygons that generate the points), you will get a small bump for each polygon, which does not amount to a smoothed surface.

No exact guidelines are available for how to choose a search radius, but sometimes you can use a behavioral theory or craft your own guideline for the case at hand. For example, areas of high crime concentration often run the length of the main street through a commercial corridor and extend one block on either side. In that case, use a search radius of one city block's length.

Study the project location

1. Open **Tutorial10-2.aprx** from **Chapter10\Tutorials** and save it as **Tutorial 10-2YourName.aprx**.

The map for this tutorial shows points from a five-year study of the number of heart attacks (myocardial infarctions) that occurred in Pittsburgh outside of hospitals. These are referred to as outside of hospital cardiac attacks (OHCA). These points count the incidents by city block and are located at the block centroid. One of the authors of this book studied this data to identify public locations for defibrillators, devices that deliver an electrical shock to revive heart attack victims. One of the location criteria was that the devices be in or near commercial areas. As a result, this tutorial incorporates buffers that are commercially zoned areas, plus about two blocks (600 feet) of surrounding areas.

KDS is an ideal method for estimating the demand surface for a service or good because its data smoothing represents the uncertainty in locations for future demand, relative to historical demand. Also, heart attacks, of course, do not occur at block centroids, so KDS distributes heart attack data across a wider area.

Run KDS

Blocks in Pittsburgh average about 300 feet per side in length. Suppose that healthcare analysts estimate that a defibrillator with public access can be identified for residents and retrieved for use as far away as two and a half blocks from the location of a heart attack victim. Thus, they recommend looking at areas that are five blocks by five blocks in size (approximately 25 blocks, or 0.08 square miles), 1,500 feet on a side, with defibrillators located in the center. With this estimate in mind, you will use a 1,500-foot search radius to include data within reach of a defibrillator, plus data beyond reach to strengthen estimates.

The objective is to determine whether Pittsburgh has areas that are roughly 25 blocks in area and whether those areas, as an additional criterion specified by policymakers, have an average of about five or more heart attacks per year outside of hospitals.

1. Search for and open the **Kernel Density** tool.

2. In the **Kernel Density** tool, click the **Environments** tab. For **Cell Size**, type **50**, and for **Mask**, select **Pittsburgh**.

Environment settings made in a tool apply to running only the tool itself and not to other tools, such as when you set environments on the **Analysis** tab.

3. In the **Kernel Density** tool, click the **Parameters** tab and apply these settings:
   - **Input point or polyline features**: OHCA
   - **Population field**: TOTAL
   - **Output raster**: HeartAttackDensity
   - **Search radius**: 1500

4. Turn off the **OHCA** and **Commercial Area Buffer** layers to see the smoothed layer.

   The default symbolization is not effective in this case, so you will resymbolize the layer.

5. Symbolize the layer:
   - **Primary Symbology**: Classify
   - **Method**: Standard Deviation
   - **Interval Size**: 1/4 standard deviation
   - **Color Scheme**: Condition Number (colors include green to yellow to red)

6. Close the **Symbology** pane.

   The smoothed surface provides a good visualization of the **OHCA** data, whereas the original **OHCA** points, even with graduated point symbols for number of heart attacks, are difficult to interpret. Try turning the **OHCA** layer on and off to see the correspondence between the point versus smoothed data.

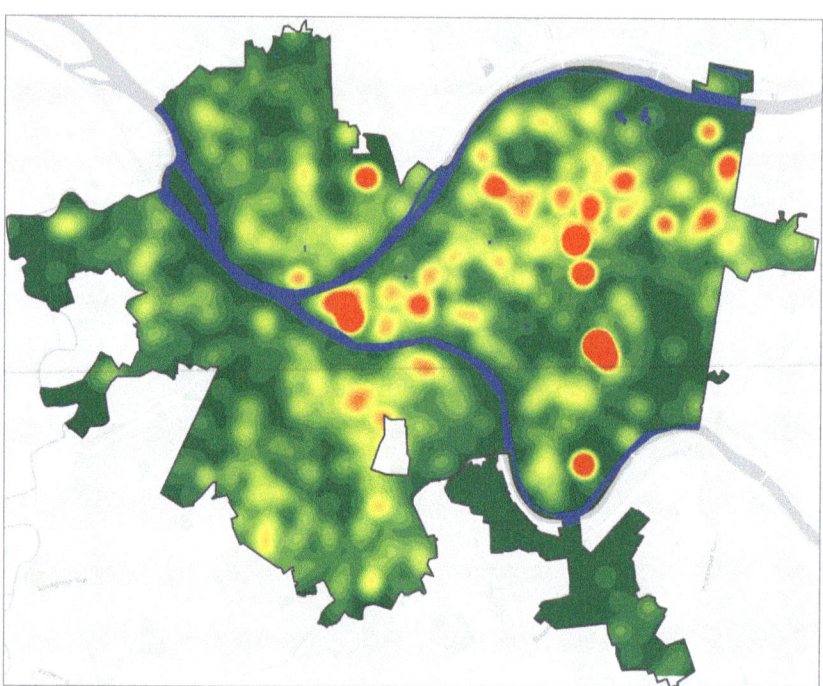

7. Turn off the **HeartAttackDensity** layer and save your project.

Create a threshold contour layer for locating a service

Assuming that target areas will be around a tenth of a square mile in area, suppose policymakers decide on a threshold of 250 or more heart attacks per square mile in five years (or 50 per square mile per year). You will create vector contours from the smoothed surface to represent this policy.

You will create only one contour, 250, for the threshold defining a target area.

1. Search for and open the **Contour List (3D Analyst Tools)** tool. Apply these settings and run the tool:
   - **Input raster**: HeartAttackDensity
   - **Output polyline features**: Threshold
   - **Contour values**: 250

2. Symbolize **Threshold** with a 2-pt bright-red color.

3. Turn off **Rivers** and turn on **Commercial Area Buffer**.

   Relatively few areas—seven—meet the criteria, and three of those areas appear quite small. All seven threshold areas are in or overlap commercial areas, so you can consider all seven as potential sites for defibrillators. Do any of the areas have the expected number of heart attacks per year to warrant defibrillators? The next section addresses this question.

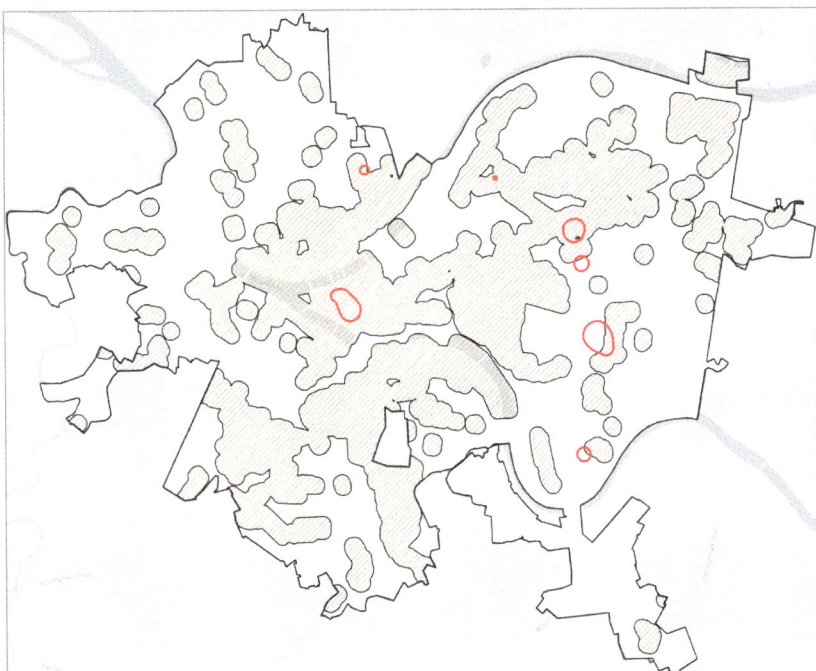

4. Turn off the **Commercial Area Buffer**.

Use threshold areas to estimate annual heart attack rates

To estimate annual heart attacks, you can select **OHCA** centroids within each threshold area, sum the corresponding number of heart attacks, and divide by five because OHCA is a five-year sample of actual heart attacks. You will use the **Threshold** boundaries in a selection by location query, in which case the **Threshold** layer must be polygons. However, if you examine the properties of the **Threshold** layer, you will see that it is a line feature class and not a polygon feature class, even though all seven areas look like polygons. The tool you ran to create **Thresholds** creates lines because some peak areas of an input raster can overlap with the border of the mask—Pittsburgh, in this case. The lines for such cases would not be closed but left open at the border. Nevertheless, ArcGIS Pro provides a tool to create polygons from lines, which you will use.

1. Search for and open the **Feature to Polygon** tool. Apply these settings and run the tool:
    - **Input Features**: Threshold
    - **Output Feature Class**: ThresholdAreas

2. Open the **ThresholdAreas** attribute table and take note of the number of polygons.

3. Search for and open the **Summarize Within (Analysis Tool)** tool. Apply these settings and run the tool:
   - **Input Polygons**: ThresholdAreas
   - **Input Summary Features**: OHCA
   - **Output Features**: **SummarizedAreas**
   - **Keep all input polygons** check box: unchecked
   - **Field**: TOTAL
   - **Statistic**: Sum

4. Open the **SummarizedAreas** attribute table and sort by **Sum TOTAL** in descending order.

One of the **ThresholdAreas** polygons had no **OHCA** points inside the polygon but was surrounded by points that contributed to its peak density. Unchecking the **Keep all input polygons** box eliminated that polygon from being summarized. The best candidate for a defibrillator has an annual average of 104/5 = 20.8 heart attacks per year and is not in the central business district (CBD), the area where the westernmost threshold area lies, but is the large area on the east side of Pittsburgh. The CBD is in the second row with an average of 68/5 = 13.6 heart attacks per year. The last row's area is the only one that does not meet the criterion of at least five heart attacks per year on average, with 4.2.

OBJECTID *	Shape *	Shape_Length	Shape_Area	Sum TOTAL	Count of Points
2	Polygon	7814.758118	4603564.31039	104	11
3	Polygon	7441.541291	3748676.979634	68	26
5	Polygon	5486.38949	2384222.400387	45	9
1	Polygon	3290.828573	860602.67515	27	2
4	Polygon	3579.388387	1018593.52118	26	2
7	Polygon	2183.844181	378193.007556	21	2

5. In the **ThresholdAreas** attribute table, select the polygon that was not included in the **SummarizedAreas** table.

The **ThresholdAreas** polygon that has no **OHCA** points was not included in the previous table. The polygon is about a city block in size and is predicted to be in a peak density area based on contributions from the kernels of nearby **OHCA** points, but it has no points itself. Considering the polygon's small size, its 28 nearby heart attacks in the five-year sample, and its distance within five blocks of the peak area, perhaps the polygon also warrants consideration as a defibrillator site.

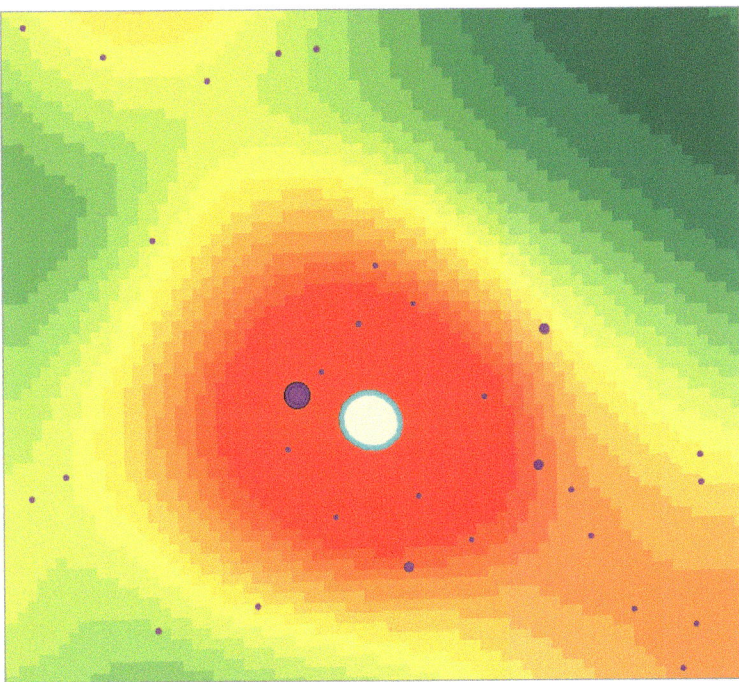

6. Save your project.

Tutorial 10-3: Building a risk index model

In this tutorial, you will learn more about creating and processing raster layers using KDS. You are also introduced to **ModelBuilder** in ArcGIS Pro, which is used for building models. Instead of writing computer code, you can drag tools to a canvas (an editing interface) and connect them in a workflow using **ModelBuilder**. ArcGIS Pro writes the script for your model in the Python scripting language. Ultimately, you can run your model as you would any other ArcGIS Pro tool.

The model you will build in this tutorial calculates an index for identifying poverty areas of a city by analyzing rasters for the following poverty indicators:
- Number of households below the poverty level
- Female-headed households with children under the age of 18
- Population 25 or older with less than a high school education
- Number of workforce males ages 16 to 64 who are unemployed

Low income alone, as defined by the US Census Bureau, may not be enough to identify poverty areas of interest for policymakers or other stakeholders. For example, female-headed households with children are among the poorest of the poor, so these populations may merit increased consideration. Likewise, populations with low educational attainment or low employment levels may be favored targets of governmental programs. Although related to poverty, each of the four indicators may have related but varying spatial distributions and maps.

Dawes (1979) provides a simple method for combining indicator measures into an overall index. If you have a reasonable theory that several variables are predictive of a dependent variable of interest such as poverty (whether the dependent variable is observable), Dawes contends that you can proceed by removing scale from each input variable and averaging the scaled inputs to create a predictive index. Alternatively, you can assign different weights to different variables according to your preferences. A good way to remove scale from a variable is to calculate z-scores, subtracting the mean and dividing by the standard deviation for each value of a variable. Each standardized variable has a mean of zero and standard deviation of one (and therefore no scale).

Table 10-1 has the means and standard deviations needed to calculate z-scores for the poverty indicators in Pittsburgh. The input raw data is annual averages by block group from the 2015 through 2019 American Community Surveys as provided by the Census Bureau. If you use the raw data instead of z-scores, households below the poverty level, considering their high mean of 71.7, have more than 50 percent weight in an average index than female-headed households with children and population with low education attainment. Therefore, using raw data instead of standardized data would inadvertently assign roughly twice as much weight to households below the poverty level than other predictive variables.

Table 10-1: Means and standard deviations for calculating z-scores

Indicator variable	Mean	Standard deviation
Female-headed households with children population	45.2	66.8
Less than high school education population	41.5	42.2
Male unemployed population	54.0	68.1
Below poverty level number of households	71.7	74.5

The workflow to create the poverty index has three steps:
1. Calculate z-scores for each of the four indicators.
2. Create kernel density maps for all four z-score variables.
3. Use a tool to average the four weighted raster surfaces and create the index raster layer.

Experts and stakeholders in the policy area using the raster index can judgmentally give more weight to some variables than others if they choose. They must use only nonnegative weights that sum to 100. If judgmental weights for four input variables are 70, 10, 10, and 10, the first variable is seven times more important than each of the other three variables. With different stakeholders possibly having different preferences, having a macro allows you to repeatedly run the macro with different sets of weights for the multiple-step process for creating an index raster layer. For example, some policymakers (educators and grant-making foundations) may want to emphasize unemployment or education and give those inputs more weight than others, whereas other stakeholders (human services professionals) may want to heavily weight female-headed households.

You need to standardize the input variables only once, so you will do step 1 manually, but you will complete steps 2 and 3 of the workflow using a **ModelBuilder** model for creating indexes.

Suppose that you have determined a KDS search radius that best represents the spatial distribution of input variables for a predictive index. If the spatial distribution of the input points has gaps of a certain kind, then, as explained in this paragraph, you will need to adjust the standardized values used in creating a predictive index. One example of such gaps is reported serious crime incidents: Parts of a study area may be free of such crimes, leading to spatial gaps in KDS. Another example is tornadoes. Large parts of a study area may not have tornadoes. To determine whether you have gaps that cause problems in KDS, create a dissolved buffer around all input points to KDS with a radius the size of the KDS search radius. The problem is that the KDS algorithm assigns zero to the smoothed output surface in all such gap areas. Zero, however, is the mean of each standardized variable, so the estimated surface would be at the mean in gaps instead of being the minimum density of the entire smoother surface. One remedy for this problem is to sort the standardized input variables in ascending order and determine the smallest positive number or integer that, when added to the input predictive variables, makes them all positive. Then add the number or integer to all standardized inputs for use in KDS so that the zero assigned by KDS in gaps is the minimum. The block group data for this tutorial has only very small gaps, so no modification is needed.

Furthermore, the standardized inputs to other kinds of spatial gaps may not need any modification. For example, if input points are center points of polygons (for example, center points of block groups), some large-size polygons may create gaps relative to the KDS search radius. In this case, the zero mean assigned by the KDS algorithm to these gaps may be an acceptable estimate of the inputs. The mean of a variable is the best estimate of a randomly selected observation.

Explore the project area

1. Open **Tutorial10-3.aprx** from **Chapter10\Tutorials** and save it as **Tutorial 10-3YourName.aprx**.

   This map has block group center points in a three-mile buffer of Pittsburgh. Each block group center point includes the corresponding four poverty indicators from table 10-1 in its attributes. For the data standardization and averaging process to be valid for creating an index, all input variables must come from the same point layer (in this case, block group center points). In other words, it is not valid to use block group data for some variables and census tract data for others.

You will use the three-mile buffer of the study region (Pittsburgh) instead of Pittsburgh itself for two reasons. First, KDS uses the northernmost, easternmost, southernmost, and westernmost points of its input point layer to define its map extent. If the inputs are polygon center points in a study region, the corresponding KDS raster map will be cut off and not quite cover the entire study region. The block group center points in the buffer yield KDS raster layers that extend somewhat beyond Pittsburgh's border, but the Pittsburgh mask will show only the portion within Pittsburgh. Second, in applying KDS, a buffer eliminates the boundary problem for KDS estimation caused by abruptly ending data at the city's edge. KDS estimates benefit from the additional data provided by the buffer beyond the city's edge.

> **YOUR TURN**
>
> In the following section, you will include kernel density smoothing of the **PittsburghBlkGrps** point data. The kernel density estimation will use 2,500-foot radius kernels. Create a **2500**-foot dissolved buffer for **PittsburghBlkGrps**. Confirm that only very small gaps exist in the coverage of kernels for Pittsburgh's three-mile buffer so that standardized values, with negative and positive values, will yield correct kernel density estimates. Remove the layer once you are done.

Standardize an input attribute

PittsburghBlkGrps already has three of four input attributes standardized and ready to use in the poverty index that you will compute, but the **FHHChildren** attribute has not yet been standardized. For practice purposes, you will standardize this attribute next.

1. Open the **PittsburghBlkGrps** attribute table.

2. Right-click the **FHHChildren** field and click **Summarize**.

3. In the **Summary Statistics tool**, for **Field** and **Statistic Type**, select **FHHChildren** and **Mean**, respectively. For a second **Field** and **Statistic Type**, select **FHHChildren** and **Standard deviation**, respectively.

4. Clear the **Case Fields** and run the tool.

```
Summary Statistics
Input Table
  PittsburghBlkGrps
Output Table
  PittsburghBlkGrps_Statistics
Statistics Fields
  Field        Statistic Type
    FHHChildren    Mean
  × FHHChildren    Standard deviation

Case Field
```

5. Open the output table, **PittsburghBlkGrps_Statistics**.

   Rounded to one decimal place, **MEAN_FHHChildren** is 45.2 and its standard deviation is 66.8.

6. Open the **PittsburghBlkGrps** attribute table and open the **Fields** view.

7. At the bottom of the **Fields** table, use the following settings to create a new field:
   - **Field Name**: **ZFHHChld**
   - **Data Type**: Float

8. Click **Save** and close the **Fields** view.

9. In the attribute table, right-click the **ZFHHChld** column heading and click **Calculate Field**.

10. Below the **Fields** list, after **ZFHHChld** =, create the following expression:

    ```
    (!FHHChildren! - 45.2)/66.8
    ```

11. Click **OK**.
    As a check, the **ZFHHChld** value for the record for **ObjectID 1** is −0.212575.

12. Close the tables and save your project.

Set the geoprocessing environment for raster analysis

You will set the cell size of rasters you create to 50 feet and use Pittsburgh's boundary as a mask.

1. On the **Analysis** tab, click **Environments**.

2. Under **Raster Analysis**, for **Cell Size**, type **50**, and for **Mask**, browse to **Chapter10.gdb** and select **Pittsburgh**.

3. Click **OK**.

Create a new toolbox and model

A toolbox contains models. When your project was created, ArcGIS Pro built a toolbox named **Chapter10.tbx**, which is where your model will be saved.

1. On the **Analysis** tab, in the **Geoprocessing** group, click **ModelBuilder**.

2. On the **ModelBuilder** tab, in the **Model** group, click **Properties**. Apply these settings:
    - Name: **PovertyIndex** (no space between the two words)
    - Label: **Poverty Index**

3. Click **OK**.

4. On the **ModelBuilder** tab, in the **Model** group, click **Save**.

Add processes to the model

ModelBuilder has a drag-and-drop environment: You search for tools, drag them into your model, and fill out their input, output, and parameters.

1. In the **Geoprocessing** pane, search for, but do not open, the **Kernel Density** tool.

2. Drag the **Kernel Density** tool into your model.

3. Drag the tool and its output to the upper center of the model window.

4. Repeat the process of adding the **Kernel Density** tool three more times.

You will need a total of four kernel density processes, one for each of the four poverty indicators.

5. Search for the **Raster Calculator (Spatial Analyst Tools)** tool and drag it to the right of your other model components. Arrange your model elements as shown in the figure by dragging rectangles around elements to select them and dragging them into place.

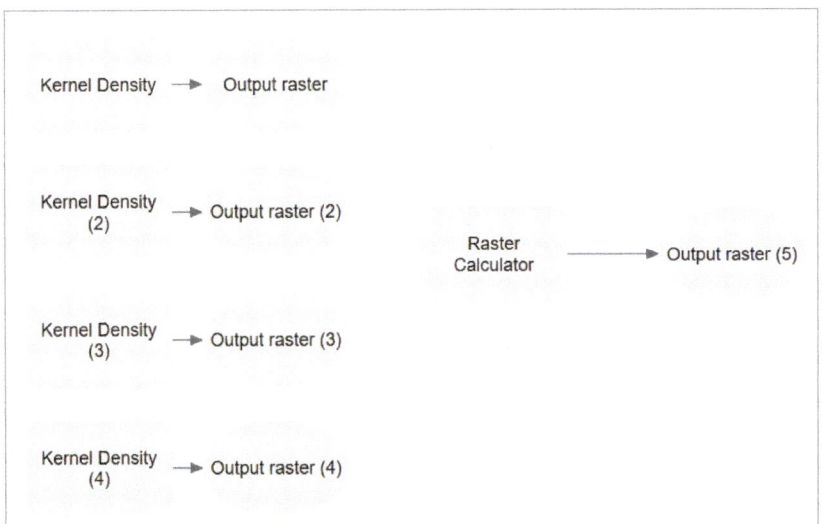

Configure a kernel density process

Once fully configured, processes and their valid inputs and outputs are filled with color.

1. Double-click the first **Kernel Density** process and apply the following settings:
   - **Input point or polyline features**: PittsburghBlkGrps
   - **Population field**: ZFHHChld
   - **Output raster**: **FHHChldDensity**
   - **Output cell size**: **50**
   - **Search radius**: **2500**
   - **Area units**: Square feet

   The 2,500-foot search radius is a judgmental estimate. This area is large enough to produce good KDS surfaces with sufficient detail at the neighborhood level.

2. Click **OK**.

3. Right-click the first **Kernel Density** process and rename it **FHHChildren Kernel Density**.

4. Save your model.

> **YOUR TURN**
>
> Configure the three remaining **Kernel Density** processes, each with **PittsburghBlkGrps** as the input, an **Output cell size** of **50**, and a **Search radius** of **2500** square feet. Refer to table 10-2 for the respective population fields to use and what to rename the process. Resize and move the model elements to improve readability.
>
> **Table 10-2: Kernel Density model data**
>
> | Population Field | Output Raster | Renamed Process |
> |---|---|---|
> | ZLessHSEdu | NoHighSchDensity | NoHighSch Kernel Density |
> | ZM16To64NotEmploy | MaleUnempDensity | MaleUnemp Kernel Density |
> | ZHseHldsBelowPov | PovertyDensity | Poverty Kernel Density |

Configure the raster calculator process

1. Double-click **Raster Calculator** and create the following expression in the **Map Algebra expression** field:

   ```
   25 * "%FHHChldDensity%" + 25 * "%NoHighSchDensity%"
   + 25 * "%MaleUnempDensity%" + 25 *
   "%PovertyDensity%"
   ```

   Tip: Double-click density rasters to enter them in the expression.

   The weights serve as default values for running the model unless users change them.

2. For **Output raster**, type **PovertyIndex**.

3. Click **OK**.

   The numbers of the blue input block groups do not need to match those in the figure. **ModelBuilder** just needs the names of all model elements to be unique. Review all four KDS processes by double-clicking each one to make sure they have the correct parameters. Now, the output raster layers from the four kernel density processes are inputs to the **Raster Calculator** tool, and all model elements have color fill and are ready to be run.

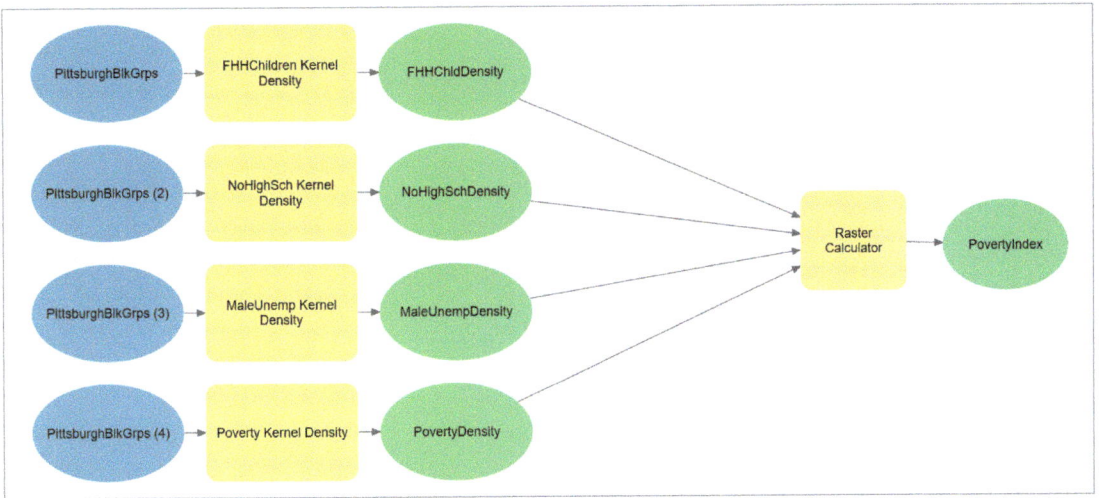

4. Save the model, but do not run it yet.

Run a model in edit mode

When you build a model, you can run the entire model by clicking its **Run** button, which you will do next. However, you can run each process one at a time, following the workflow by right-clicking processes and clicking **Run**, which allows you to isolate and fix errors.

1. On the **ModelBuilder** tab, in the **Run** group, click **Run**.

   The model runs and produces a log. If there is an error, the log provides information about it. If errors occur, you can fix and validate the model by clicking the **ModelBuilder** tab and clicking **Validate** in the **Run** group. When the validation is complete, you can rerun the model.

2. Review and close the log file.

3. In the model, right-click **PovertyIndex** and click **Add To Display**.

4. Open your map and turn off **PittsburghBlkGrps**.

You will improve symbolization for **PovertyIndex**.

Symbolize a KDS raster layer and save its layer file

Symbolizing raster layers can include many categories to represent continuous surfaces. You will use standard deviations to create categories. Finally, you will save your symbolization as a layer file for automatic use whenever the model creates its output, **PovertyIndex**, with whatever weights the user chooses.

1. Open the **Symbology** pane for **PovertyIndex:PovertyIndex** and apply these settings:
   - **Primary Symbology**: Classify
   - **Method**: Standard Deviation
   - **Interval Size**: 1/4 standard deviation
   - **Color Scheme**: Condition Number (colors include green to yellow to red)

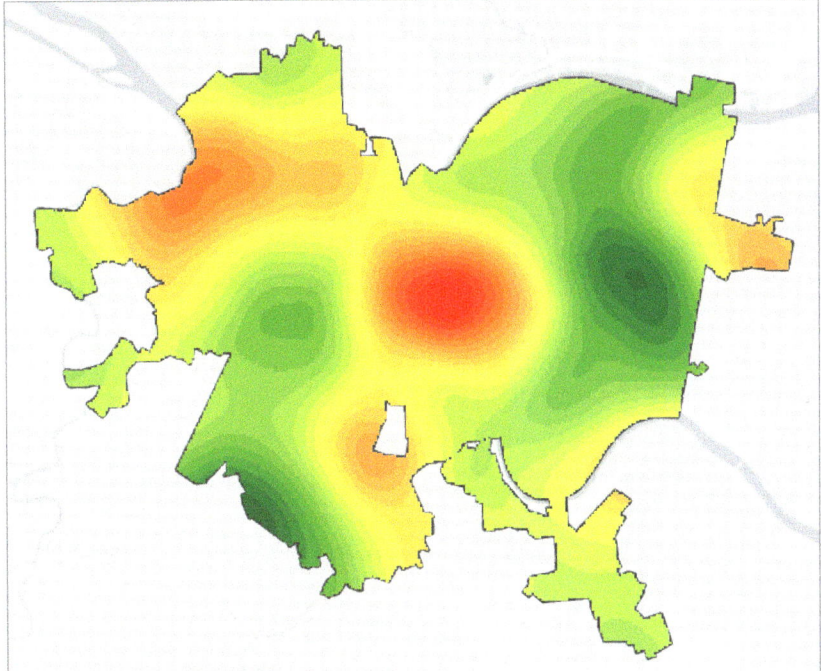

2. Close the **Symbology** pane.

   There is an intensely poor area in the center of Pittsburgh with additional poor areas.

3. In the **Contents** pane, right-click **PovertyIndex:PovertyIndex**, click **Sharing > Save As Layer File**, and save as **PovertyIndex.lyrx** to **Chapter10\Tutorials**.

   You will use the layer file to symbolize **PovertyIndex** automatically in future runs of the model.

4. Remove **PovertyIndex:PovertyIndex** from the **Contents** pane and save your project.

Add variables to the model

An objective for the **PovertyIndex** model is to allow users to change the poverty index's weights. To accomplish this change, you must create variables that will store the weights that users enter. Then you designate each variable as a parameter for user input. **ModelBuilder** automatically creates a user interface for your model, just like the interface for any tool. Users can then enter weights for your model in the interface as an alternative to the default equal weights of 25 each for the poverty index model.

1. Return to the **PovertyIndex** model.

   Each process and output has a drop shadow, indicating that it is finished running.

2. On the **ModelBuilder** tab, in the **Run** group, click **Validate**.

   The **Validate** function places all model components in the state of readiness to edit or run again. The drop shadows disappear.

3. On the **ModelBuilder** tab, in the **Insert** group, click **Variable**. For **Variable Data Type**, select **Variant** and click **OK**. Move the **Variant** variable in the area above the **Raster Calculator** tool.

   ArcGIS Pro can determine the actual data type of a variant data type variable from entered values. For example, when you give the variable the value 25, ArcGIS Pro will treat the value as an integer variable.

4. Right-click **Variant**, click **Open**, type **25**, and click **OK**.

   The value of 25 is the default value for running the model if users do not change it.

5. Right-click the **Variant** variable, click **Rename**, type **FHHChldWeight**, and press **Enter**.

6. Right-click **FHHChldWeight** and click **Parameter**.

   This action places a letter *P* near the variable, making the variable a parameter. In other models you build, consider making the model inputs (in this case, the four raster layers) parameters so users can change them when they run a tool.

> **YOUR TURN**
>
> Add three more **Variant** variables, all with value **25**, and named **NoHighSchWeight**, **MaleUnempWeight**, and **PovertyWeight**. Make each variable a parameter.

Use in-line variable substitution

In this section, you will transfer the weights stored in variables (values that will be input by users) to parameters of the **Raster Calculator** tool. The mechanism is called in-line substitution because the variables' values are substituted for the model's parameter values.

1. Open **Raster Calculator** in the model.

2. Select and delete the first **25** in the expression and replace it with **"%FHHChldWeight%"**. Be sure to remove the double quotation marks.

3. Likewise, select and replace the 25 weights for the remaining rasters, yielding the finished expression as follows:

   ```
   %FHHChldWeight% * "%FHHChldDensity%" +
   %NoHighSchWeight% * "%NoHighSchDensity%" +
   %MaleUnempWeight% * "%MaleUnempDensity%" +
   %PovertyWeight% * "%PovertyDensity%"
   ```

4. Click OK.

   Raster Calculator and its output will have color fill if your typed expression is correct. If there is no color fill, review and correct your expression, paying particular attention to spellings.

5. Save your model.

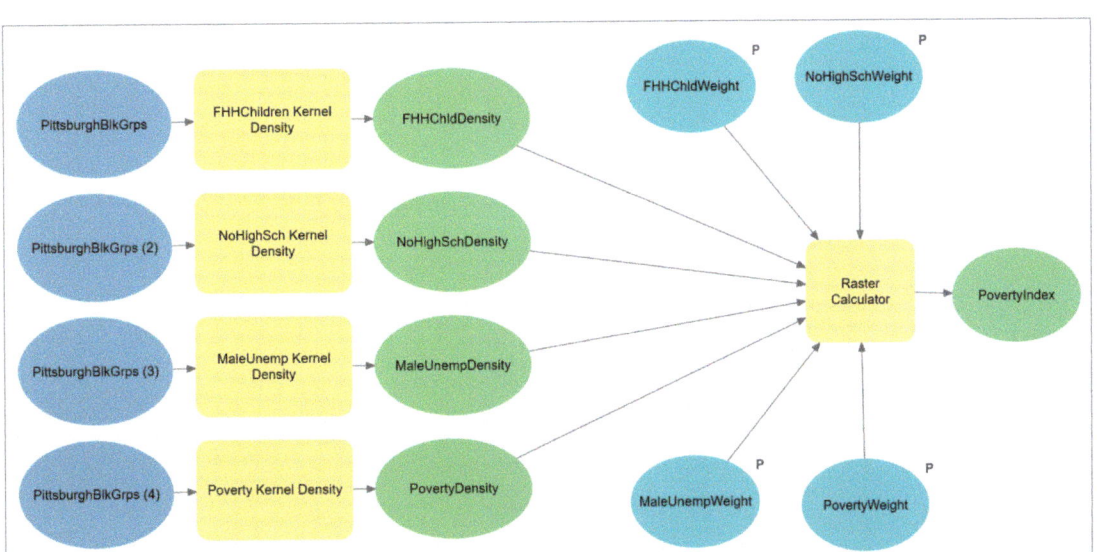

Use a layer file to automatically symbolize the raster layer when created

You will use the layer file you created earlier for the poverty index for this task. To do so, you must make the **PovertyIndex** output from **Raster Calculator** a parameter in the model.

1. Right-click **PovertyIndex** and make it a parameter.

2. On the **ModelBuilder** tab, in the **Model** group, click **Properties** and click the **Parameters** tab.

3. Locate the **PovertyIndex** row and scroll right to the end of the table. Click the **Symbology** column cell for the row.

4. In the editable text field for that cell, click **Browse**, browse to **Chapter10\Tutorials**, and double-click **PovertyIndex.lyrx**.

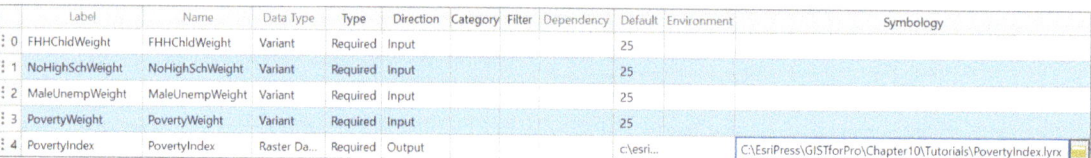

5. Click **OK**.

6. Save and close your model.

Run your model tool

Congratulations! Your model tool is ready to use.

1. With your map open, ensure that **PovertyIndex:PovertyIndex** is removed from the **Contents** pane.

2. In the **Geoprocessing** pane, search for and open the **Poverty Index** model tool.

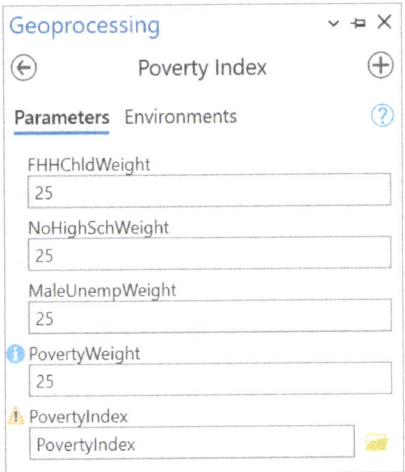

3. Leave the weights at their default settings and run the model tool.

   The model tool computes the poverty index, adds the **PovertyIndex** raster to your map, and symbolizes it with your layer file.

4. Save your project.

> **YOUR TURN**
>
> Rerun your model tool with different weights (nonnegative and sum to 100), such as **70**, **10**, **10**, and **10**, saving the output as **PovertyIndex_Chld70**. Notice how the output changes. For a class project or work for a client, you must symbolize **PovertyIndex** manually to get the best color scheme and categories for a final model, depending on the distribution of densities produced. When you finish, save your project and close ArcGIS Pro.

Assignments

This chapter has assignments to complete that you can download with data from ArcGIS Online at links.esri.com/GISTforPro3.4Assignments.

References

Dawes, Robyn M. 1979. "The Robust Beauty of Improper Linear Models in Decision Making." *American Psychologist* 34 (7): 571–82.

CHAPTER 11

3D GIS

LEARNING GOALS

- Explore global scenes.
- Learn how to navigate scenes.
- Create local scenes and TIN surfaces.
- Create z-enabled features.
- Create 3D buildings and bridges from lidar data.
- Work with 3D features.
- Use procedural rules and multipatch models.
- Create an animation.

Introduction

This chapter introduces 3D display and data and map processing. Scenes, which can be thought of as 3D maps, provide insights that aren't readily apparent from 2D visualizations of the same data. For example, instead of assuming the presence of a valley from 2D contours, in 3D you see the valley as well as the difference in height between the valley floor and a ridge. Scenes allow you to design, visualize, communicate, and analyze for better decision-making.

Scenes use one of two viewing modes. The first, global, is for large-extent, real-world content in which the curvature of the earth is important. The second, local, is for small-extent content in a projected coordinate system or for situations in which the curvature of the earth isn't important. You can switch scenes between global and local. You can also modify a scene's settings, depending on the scale of the project. Rendering scenes is slower than rendering 2D maps, and proper computer hardware and configuration are necessary.

This chapter explains procedural methods for rapidly creating 3D content. The methods describe how to construct multiple 3D models based on feature attributes—such as building heights and roof shape types—rather than creating a single, specific 3D model. Procedural rules, which define patterns, are authored in ArcGIS CityEngine®, a 3D modeling software program for urban environments. You can reuse procedural rules in other ArcGIS programs after they have been exported as rule package files.

This chapter also explains the use of lidar data and visual analysis tools, such as line of sight, and introduces 3D animation.

Tutorial 11-1: Exploring a global scene

Global scenes use a default elevation surface, WorldElevation3D/Terrain3D, from an ArcGIS Online map service and the GCS WGS84 coordinate system. In this tutorial, you will explore a global scene's properties and learn how to navigate in 3D. Advantages of the global scene include working in large extents or in multiple geographic areas.

Explore a scene's properties

1. Open **Tutorial11-1.aprx** from **Chapter11\Tutorials** and save it as **Tutorial 11-1YourName.aprx**.

   The project opens with a scene using a default elevation surface, **Terrain3D**. In the **Contents** pane, **3D Scene** is labeled with a globe icon, indicating that it's a global scene. The scene shows the city of Pittsburgh, including the central business district, North Shore, South Shore, and Mount Washington neighborhoods. There are no GIS features yet—just a basemap.

2. In the **Contents** pane, right-click **WorldElevation3D/Terrain3D** > **Properties**.

   The scene's elevation surface is different from a basemap, and understanding the elevation surface, map units, and heights is important in a scene.

3. In the **Layer Properties** window, click the **Source** tab.

   The scene's elevation source is visible, showing the elevation service name, **WorldElevation3D/Terrain3D**, its data type, and its server location in arcgis.com.

Navigate a scene with a mouse and keyboard keys

Next, you will view the map using a predefined 3D bookmark and explore it using mouse and keyboard shortcuts.

1. Zoom to the **Rivers** bookmark.

   The view shows that the terrain is higher in the Mount Washington neighborhood, on the left of the view.

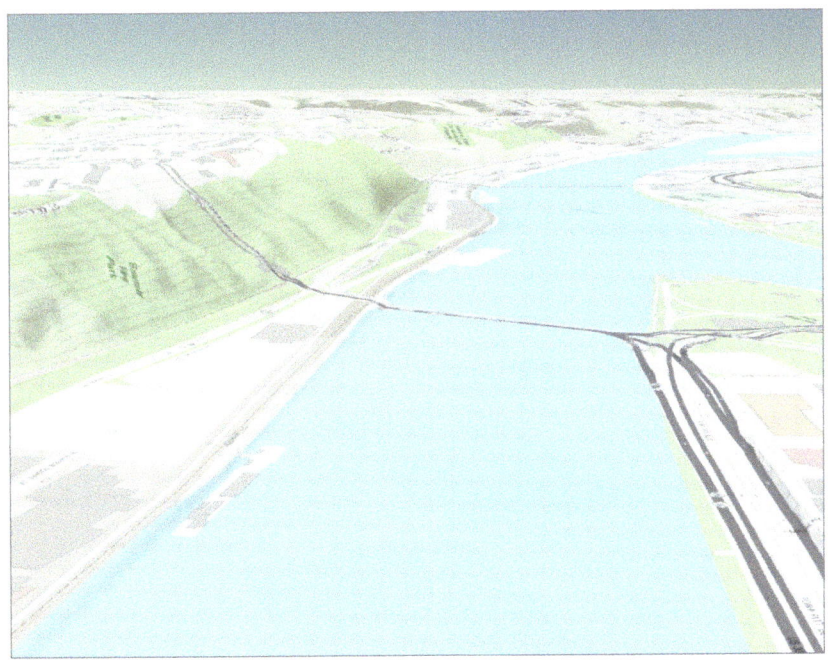

Sometimes, you can get disoriented in a 3D view, so learning a few useful shortcuts can help you return to a familiar orientation. Experiment with the following keyboard shortcuts, commonly used to manipulate a 3D view, in the next set of steps.

2. Drag the mouse wheel button to adjust (tilt) the view.

3. On the keyboard, press the **J** key to move the map down or the **U** key to move the map up.

4. Press the **A** or **D** key to rotate the view clockwise or counterclockwise.

5. Press the **W** or **S** key to tilt the camera up and down.

6. Press the **left**, **right**, **up**, or **down arrow** keys to move the view.

7. Press the **B** key and use the **left** mouse button or **arrow** keys to look around your view.

8. Press the **N** key to view true north.

9. Press the **P** key to look straight down at your map.

Change the basemap

You can display various basemaps with the current surface elevation. If you want to see imagery details, they will be draped to the elevation surface.

1. Zoom to the **Football Stadium** bookmark.

2. Change the basemap to **Imagery** and zoom out a few times if necessary to see the Heinz Field football stadium along the river.

   The imagery drapes to the elevation surface, showing the football stadium, rivers, and trees along the hills above Pittsburgh's South Shore.

3. Zoom to the **Baseball Stadium** bookmark, zoom out, and use the mouse or keyboard to see additional views.

4. Change the basemap back to **Topographic** and zoom to the **Rivers** bookmark.

> **YOUR TURN**
>
> Explore another geographic area, perhaps your hometown, your favorite vacation spot, or a city or area you have always wanted to visit.

Exaggerate and apply a shade and time to a surface

Sometimes, subtle or important changes in the landscape can be emphasized by adding visual effects to the layer. For example, you can graphically exaggerate the height of a mountainous area to help it stand out. This exaggeration doesn't change the elevation but visually makes features more prominent. Another effect includes adding lighting or illumination sources through shading or time of day.

1. In the **Contents** pane, under **Elevation Surfaces**, click **Ground** and click the **Elevation Surface Layer** tab.

2. For **Vertical Exaggeration**, type **3.00**, and check the box for **Shade Relative to Light Position**.

3. In the **Contents** pane, right-click **3D Scene** and click **Properties**.

4. In the **Map Properties** dialog box, click the **Illumination** tab. Under **Illumination defined by**, click **Date and time** and click **OK**.

The elevation will be exaggerated, and shadows will depend on your current date and time.

> **YOUR TURN**
>
> Pan and navigate the scene to see the exaggeration from different views. Navigate to another area you know is mountainous. Save your project.

Tutorial 11-2: Creating a local scene and TIN surface

Advantages of local scenes include using your own elevation surface data such as triangulated irregular networks (TINs) or lidar data, using a projected coordinate system, managing features below a surface (for example, subways or waterlines), and perhaps more accessibility to edit data. You can also set the coordinate system for a local scene to local coordinates (for example, state plane) and use the surface offline. In this tutorial, you will create a TIN surface from contours, change its symbology, and use it as the elevation surface in a local scene.

Set a local scene

1. Open **Tutorial11-2.aprx** from **Chapter11\Tutorials** and save it as **Tutorial 11-2YourName.aprx**.

    The project opens with a scene named **TIN Surface Scene** with 2D layer contours, street curbs, parks, and rivers draped to the default **Terrain3D** elevation surface in a global scene. The basemap is **Light Gray Canvas Base**, covering the entire earth. You will convert the global scene to a local scene and clip the basemap to the study area. Unlike the geoprocessing **Clip** tool, this process clips the layers for display purposes only.

2. On the **View** tab, in the **View** group, click **Local**.

    The scene switches from a global to a local scene, and the icon is updated in the **Contents** pane and the view. Next, you will clip the base layer to the study area extent using the **Contours** layer.

3. In the **Contents** pane, right-click **TIN Surface Scene**. Click **Properties** and click the **Clip Layers** tab.

4. Click the down arrow and choose **Clip to a custom extent**, and for **Get extent from**, click **Contours**.

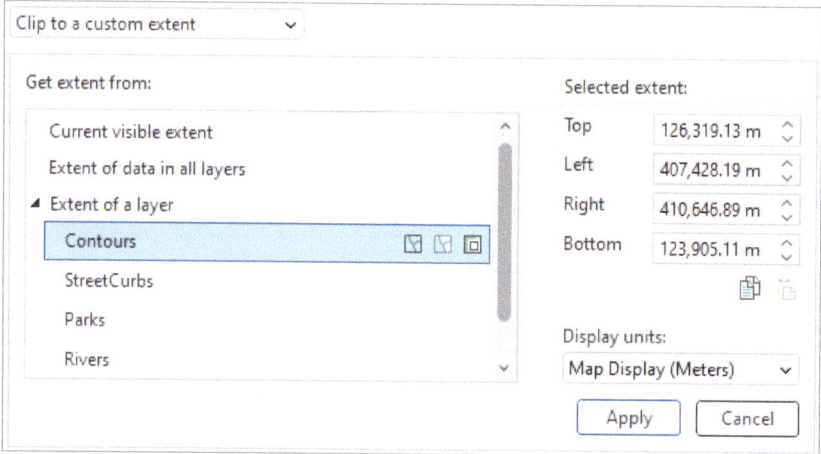

5. Click **Apply** and click **OK**.

6. Zoom to the **3D View** bookmark.

   The basemap will now display only to the extent of the **Contours** layer.

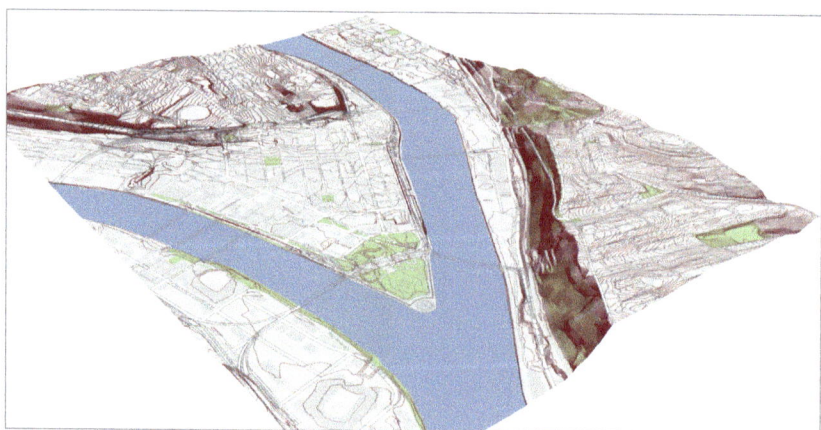

Create a TIN surface

A TIN surface is a vector data model consisting of irregularly distributed nodes and lines that are formed from x-, y-, and z-values and arranged in a network of triangles that share edges. TINs are typically used for the high-precision modeling of small areas, such as engineering applications, in which they allow calculations of surface area and volume. TIN surfaces are also useful for viewing underground features or utilities. Here, you will create a TIN from a contour layer.

1. Search for and open the **Create TIN** tool. Apply these settings and run the tool:
   - **Output TIN**: **PGH_TIN**
   - **Coordinate System**: Current Map [TIN Surface Scene]
   - **Input Features**: Contours

2. In the **Contents** pane, turn off **Contours** and the basemap and expand **PGH_TIN**.

   The TIN elevation is displayed with elevation heights from high to low.

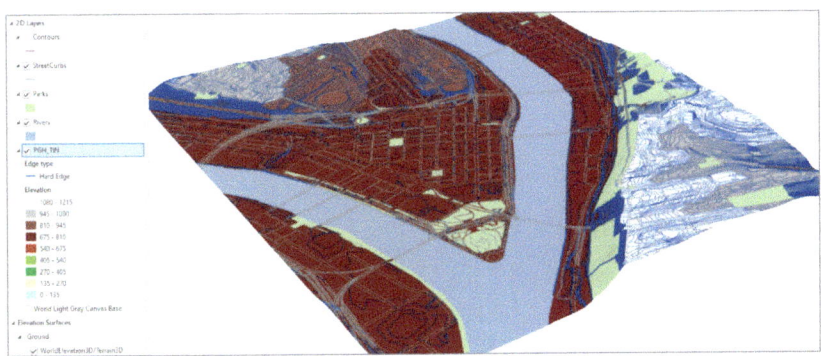

Change the scene's surface and coordinate system

The local TIN surface can now replace the surface assigned from the map service.

1. In the **Contents** pane, under **Elevation Surfaces**, right-click **Ground** and click **Add Elevation Source Layer**. Browse to **Chapter11\Tutorials**, click **PGH_TIN**, and click **OK**.

   The dataset **PGH_TIN** is the first surface listed as an elevation source.

2. Under **Elevation Surfaces**, remove **WorldElevation3D/Terrain3D**.

   This step removes this surface as an elevation source. Wait for the scene to redraw its features. The scene is now set to local coordinates, with **PGH_TIN** as the surface data, which can be used offline.

> **YOUR TURN**
>
> Change the scene's elevation units to feet, which is the default unit for NAD 1983 State Plane Pennsylvania South. Do this by opening the **Map Properties** window for **TIN Surface Scene**. On the **General** tab, change **Elevation Units** to **Feet**.

Change the symbology of a TIN

You can change the symbology of a TIN to better reflect features that the surface model represents. You will add the contour and slope symbology renderers.

1. In the **Contents** pane, under **2D Layers**, right-click **PGH_TIN** and click **Symbology**.

2. Make sure you are on the **Edges** tab.

3. For **Draw Using**, uncheck the box.

4. At the top, click **Symbolize your layer using a surface**.

5. For **Draw Using**, select **Simple**.

6. Click the current symbol, and under **Gallery**, click the first instance of **Land**.

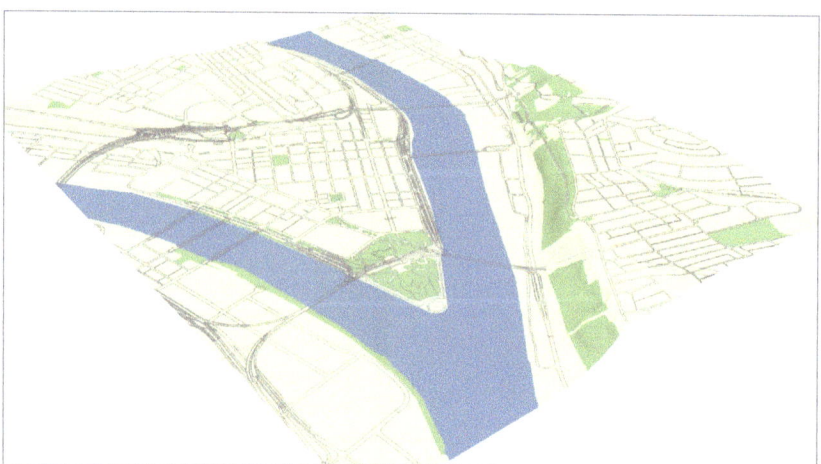

7. Zoom and pan to better see the features from different angles. Save your project.

Tutorial 11-3: Creating z-enabled features

You can create 3D content in different ways, and the corresponding workflows depend on the type of features you create. In addition to creating 3D features from scratch, you can import 3D models and symbolize 2D features as 3D features. You can also specify the source of your z-values when you create features. In ArcGIS Pro, the **Current Z Control** tool is used to set the 3D elevation source for drawing or obtaining z-values. This option is useful if more than one source is defined for a global or local scene or if you have another source not already included in the map.

In this tutorial, you'll create a 3D feature class that is z-enabled. Then you'll use the **Current Z Control** tool to set the elevation source for populating z-values. **Current Z Control** has two modes, constant and surface. Constant is used to create 3D features at an absolute height by typing an exact value—for example, a plane flying at a constant altitude. Surface uses z-values from the active elevation source you choose.

Create a z-enabled feature class for trees

1. Open **Tutorial11-3.aprx** from **Chapter11\Tutorials** and save it as **Tutorial 11-3YourName.aprx**.

   The map opens with **3D Trees Scene**, a local scene using **WorldElevation3D/Terrain3D** as the elevation surface whose extent is clipped to the **Rivers** layer. **3D Trees Scene** has the **Parks** layer draped on the surface. The scene shows a large park at the confluence of Pittsburgh's three rivers.

   To take advantage of certain 3D editing capabilities, you must ensure that an output feature class will have z-values. The **Create Feature Class** geoprocessing tool allows you to determine these settings. Here, you'll create an empty 3D feature class, digitize new features, and populate z-values directly from the map's surface. If you need to confirm whether a layer is z-enabled, you can verify the data source information listed on the **Source** page in the **Layer Properties** dialog box.

2. Search for and open the **Create Feature Class** tool. Apply these settings and run the tool:
   - **Feature Class Name:** **ParkTrees**
   - **Geometry Type:** Point
   - **Has Z:** Yes
   - **Coordinate System:** Parks (NA_1983_StatePlane_Pennsylvania_South)

Digitize trees on surfaces using Z Mode

1. On the **Edit** tab, in the **Elevation** group, click the **Z Control** button (**Z Mode**).

   This step turns on the **Z Control** option.

2. Click the **Get Z from View** (cross hair) button.

3. Click a point inside the center of Point State Park (see the figure) to set the z-value (height).

   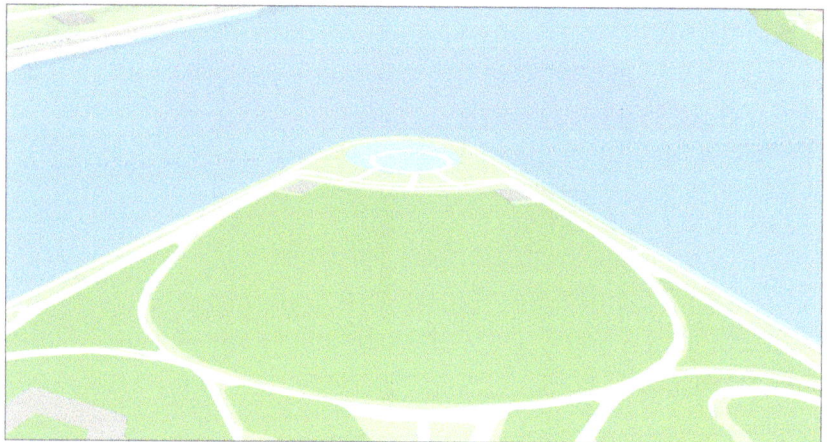

   Depending on where you click, the elevation height is about 723 feet.

4. On the **Edit** tab, in the **Features** group, click **Create**.

5. In the **Create Features** pane, click **ParkTrees**.

6. Click about eight total points to digitize trees on the left and right sides of the park.

   It may take a moment to see each point as you digitize it.

7. Save your edits and clear the selection.

Display trees as a realistic genus (type)

You can display 3D tree symbols as low-resolution, high-resolution, or thematic trees. Depending on the number of features and map purpose, you will want to experiment with all three tree types.

1. Zoom in close to the trees.

2. In the **Contents** pane, right-click **ParkTrees** and click **Properties**. On the **Display** tab, under **Display symbols in scene**, select **Real-world units** and click **OK**.

3. In the **Contents** pane, click the symbol for **ParkTrees**.

4. In the **Symbology** pane, click the symbol.

5. On the **Gallery** tab, type **Trees** in the search box and press **Enter**.

6. Scroll to **3D Vegetation – Thematic** and click **Norway Maple**.

   Next, you'll enter an exact height for the park trees.

7. Click the **Properties** tab. Under **Appearance**, for **Size**, type **20** m and click **Apply**.

8. In the **Contents** pane, right-click **3D Trees Scene** and click **Properties**.

9. In the **Map Properties** dialog box, click the **Illumination** tab, confirm that the **Display Shadows in 3D** box is checked, and click **OK**.

10. On the **Edit** tab, click the **Z Control** button (**Z Mode**) to deselect it. Save your project.

Tutorial 11-4: Creating features and line-of-sight analysis using lidar data

Lidar uses pulsed laser light from aircraft or drones or other sources to provide detailed elevation data and classification of land cover that you can use to create 2D surfaces and 3D features. Geographic lidar data is commonly available as LAS files, the industry standard of the American Society for Photogrammetry and Remote Sensing. In this tutorial, LAS files were provided by Pictometry International Corporation for a study area of Allegheny County, Pennsylvania.

The generation of 3D buildings from lidar LAS datasets requires two surface models, a digital surface model (DSM) and a digital terrain model (DTM), to create a normalized surface (nDSM), which is the difference between the DSM and DTM surfaces used to calculate building heights. The nDSM is applied to random points that are created for 2D building footprints. These footprints are used to generate z-values (heights) for each random point. The z-value of the highest point is the building height. The process finishes by creating a statistics table, which selects the maximum z-value of the random points, that is then joined to 2D building footprints, allowing for buildings to be extruded using that value. You can also use lidar data to determine line-of-sight obstructions between features, such as buildings as shown in the final section of this tutorial.

Create a LAS dataset

1. Open **Tutorial11-4.aprx** from the **Chapter11\Tutorials** folder and save it as **Tutorial11-4YourName.aprx**.

   The scene displays 2D layers of building footprints, a **World Light Gray** basemap, and a lidar delivery area used to clip the basemap. There is no height value in the building attribute table, so you can display buildings as only flat 2D polygons for now. Additional 3D layers of observer points (used for line-of-sight analysis) and bridges are turned off.

   LAS files have points classified as bare earth, vegetation, buildings, and so on, which you can view in 3D or make into raster layers. Originally, lidar data was delivered only in American Standard Code for Information Interchange (ACSII) format. With the massive size of lidar data collections, LAS was soon adopted to manage and standardize the way in which lidar data was organized and disseminated. A LAS dataset, created from original LAS data, provides fast access to lidar data without the need for data conversion to work with LAS files for a specific study area.

2. Search for and open the **Create LAS Dataset** tool. Apply the following settings and run the tool:
   - Input Files: Browse to **Chapter 11 > Data > LASFiles** and import the six LAS files.
   - Output LAS Dataset: **Chapter11LasDataset.lasd**
   - Coordinate System: Bldgs (NAD_1983_StatePlane_Pennsylvania_South)

The lidar data values are clearly shown as points and their values. You can symbolize the LAS dataset using elevations, slope, aspect, and so on. Pittsburgh's tallest building is the US Steel building, the triangular building on the right side of the study area.

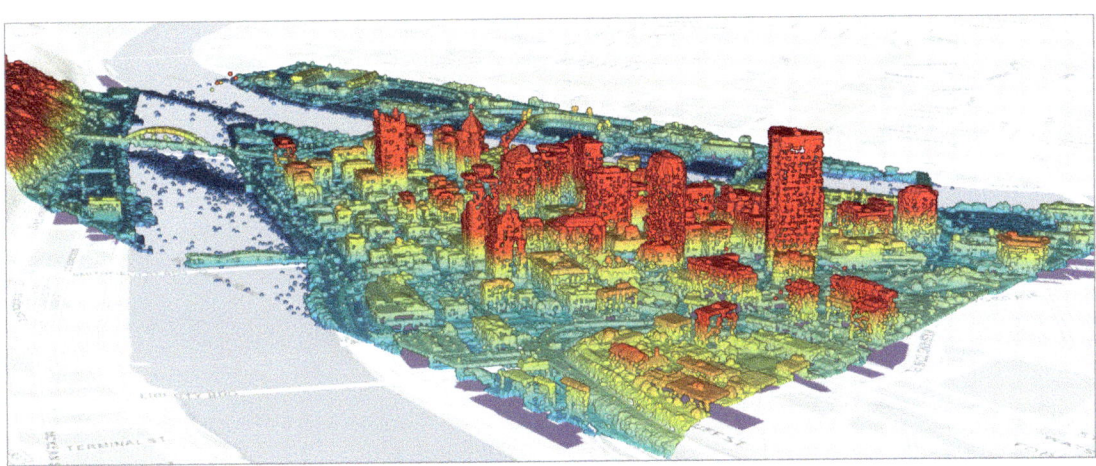

3. Explore the 3D map from various locations and zoom to the **3D View** bookmark.

Generate a raster DSM

DSMs represent the surface of the earth, including buildings, tree canopies, and other things that create a surface above the terrain. Before generating the DSM raster, you will filter lidar points to save processing. You will create a DSM using an interpolation type of binning, which is faster for processing, and a maximum cell assignment to find the highest elevation point within each cell.

1. In the **Contents** pane, right-click **Chapter11LasDataset.lasd** and click **LAS Filters > 1st Return**.

2. Search for and open the **LAS Dataset to Raster** tool. Apply these settings and run the tool:
   - **Input LAS Dataset**: Chapter11LasDataset.lasd
   - **Output Raster**: DSM
   - **Interpolation Type**: Binning
   - **Cell Assignment**: Maximum
   - **Sampling Value**: 5

3. Turn off the **Chapter11LasDataset.lasd** and **Bldgs** layers.

The values of the DSM show the range of elevations from high to low.

Generate a raster DTM

Next, you will create the DTM, a bare-earth terrain surface, containing only the topology. In many cases, a DTM is the same as a digital elevation model (DEM). Before creating the raster, you will filter the ground features.

1. Turn off the **DSM** layer.

2. Turn on the **Chapter11LasDataset.lasd** layer, right-click the layer and click **LAS Filters > Ground**.

This will filter and show only the ground features used to create the DTM.

This raster uses a different interpolation type of triangulation that takes a little longer to process but better interpolates earth surface voids found in the LAS dataset.

3. Open the **LAS Dataset to Raster** tool. Apply these settings and run the tool:
    - **Input LAS Dataset**: Chapter11LasDataset.lasd
    - **Output Raster**: DTM
    - **Interpolation Type**: Triangulation
    - **Interpolation Method**: Natural Neighbor
    - **Sampling Value**: 5

Ignore any warning messages. A raster surface of the earth's features is created.

Create an nDSM raster

An nDSM surface is the difference between the DSM and the DTM surfaces that is normalized to the bare-earth surface.

1. Search for and open the **Minus (3D Analyst Tools)** tool. Apply these settings and run the tool:
    - **Input raster or constant value 1**: DSM
    - **Input raster or constant value 2**: DTM
    - **Output raster**: nDSM

You now have a raster surface that you can apply to point features used for buildings to determine their height.

Create random points for buildings

Randomly generated points are created for each building polygon, and the nDSM raster surface is applied to each random point. The point with the highest z-value will be used as the building height.

1. Turn off the **Chapter 11LasDataset** layer and turn on the **Bldgs** layer.

2. Search for and open the **Create Random Points** tool. Apply these settings and run the tool:
   - **Output Point Feature Class**: BldgRandomPoints
   - **Constraining Feature Class**: Bldgs
   - **Number of Points**: 100
   - **Minimum Allowed Distance**: 5 US Survey Feet

> **YOUR TURN**
>
> Turn off all layers except **BldgRandomPoint**s and zoom in. Click various random points on each building. Note that the 100 points for the US Steel building will have a **CID** value (unique value for each building) of 521. Every building has a unique CID value that you will use later to join to building footprints.

Add surface information to random points

You will assign z-values (height) from the nDSM raster surface to each random point using the **Add Surface Information** tool.

1. Search for and open the **Add Surface Information (3D Analyst Tools)** tool. Apply these settings and run the tool:
   - **Input Features**: BldgRandomPoints
   - **Input Surface**: nDSM
   - **Output Property**: Z check box <checked>

> **YOUR TURN**
>
> Click various random points and note that they now have z-values. The highest value for each building will be used for the building height.

Assign a maximum value (height) to random points

The **Summary Statistics** tool calculates the maximum z-value for all buildings using the building's random points. A text file is created that you will join back to buildings, using the **CID** field.

1. Search for and open the **Summary Statistics** tool. Apply these settings and run the tool:
   - **Input Table**: BldgRandomPoints
   - **Output Table**: **BldgHeights**
   - **Field**: Z
   - **Statistic Type**: Maximum
   - **Case Field**: CID

> **YOUR TURN**
>
> Open the **BldgHeights** table and sort the **MAX_Z** field in descending order. Look for **CID 521**; the tallest building will be the US Steel building. Close the table.

Join the maximum z-value (height) to building footprints and display as 3D buildings

1. In the **Contents** pane, right-click **Bldgs** and click **Joins and Relates** > **Add Join**.

2. Join **BldgHeights** to **Bldgs** using BLDG_ID for **Input Field** and **CID** for **Join Field**.

3. Turn on the **Bldgs** layer and drag it to the top of the **3D Layers** list.

4. Turn off **BldgRandomPoints**, turn on the basemap, and zoom to the **3D View** bookmark. Save your project.

These steps required a lot of processing, but you now have 3D buildings. Notice the residential buildings on the left of the view in the Mount Washington neighborhood, as opposed to the taller high-rise buildings in the central business district.

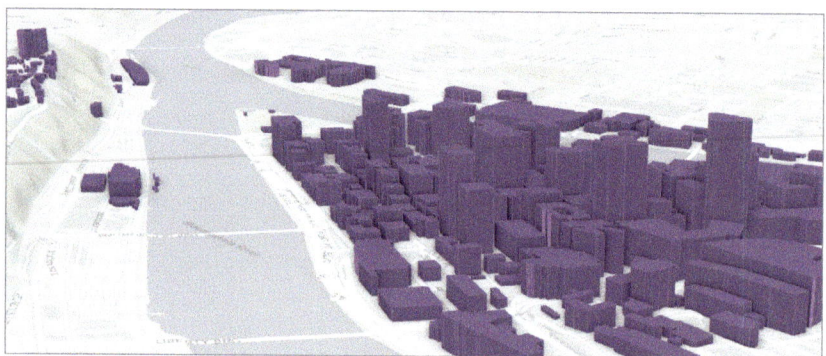

Use lidar to determine bridge elevation heights

You can view and select lidar data points to determine the elevation height to draw bridges. Pittsburgh has more than 750 bridges, but you will use data to find the height of and digitize just one.

1. Turn off the **Bldgs** layer and zoom to the **Fort Pitt Bridge** bookmark. Turn on the **Chapter11LasDataset.lasd** layer, right-click the layer, and click **LAS Filters > All Points**.

2. On the **Map** tab, click **Explore** and click various lidar points along the bridge to see the z-values.

   Points at the top of the bridge span are approximately 920 feet, whereas points at the bottom deck range from 770 feet to 800 feet. You will use 775 feet as the elevation to draw the base of the bridge.

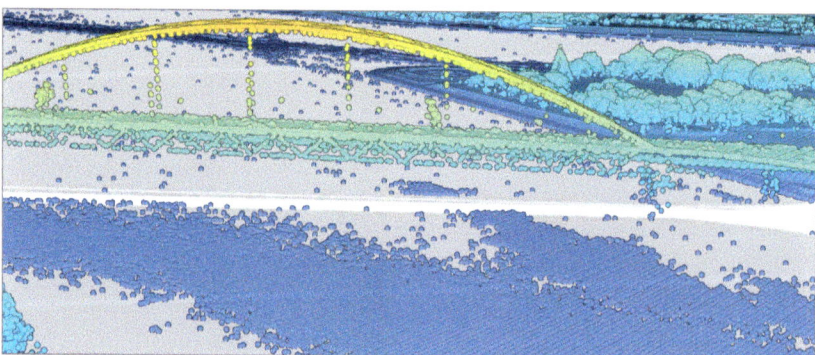

Draw a bridge using Z Mode elevation

It's easier to draw the bridge in a 2D map, and you can do so by setting the **Z Mode** elevation.

1. On the **View** tab, in the **View** group, click **Convert > To Map**. Turn on the **Bridges** layer and zoom out until you can see the bridges.

2. On the **Edit** tab, in the **Elevation** group, click the **Z Mode Constant** button and type **775** as the constant elevation.

3. In the **Features** group, click **Create**, and in the **Create Features** pane, click **Bridges > Polygon**.

4. Zoom to the bridge shown and digitize the approximate location of the bridge. Click **Finish**.

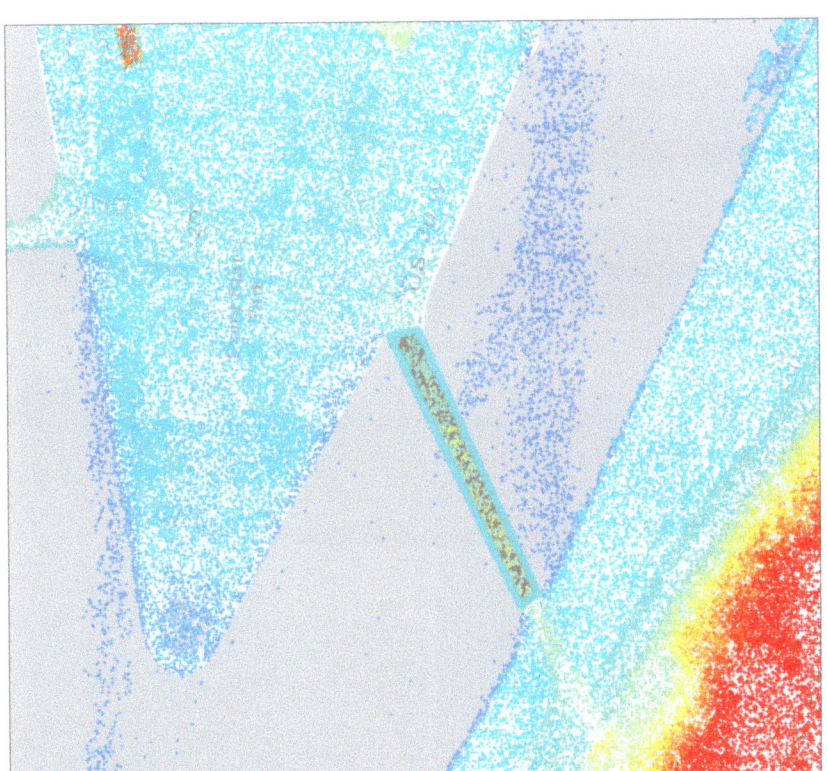

5. Save your edits and close the 2D map to return to the scene, and then turn on the **Bridges** layer.

The bottom of the bridge is at the correct elevation. You can also snap lidar points to create 3D features.

6. Click **Create** and use the **Bridges** > **Polygon** option to digitize the top of the bridge using multiple polygons. Editing a bridge span in 3D is challenging, so pan and zoom as necessary.

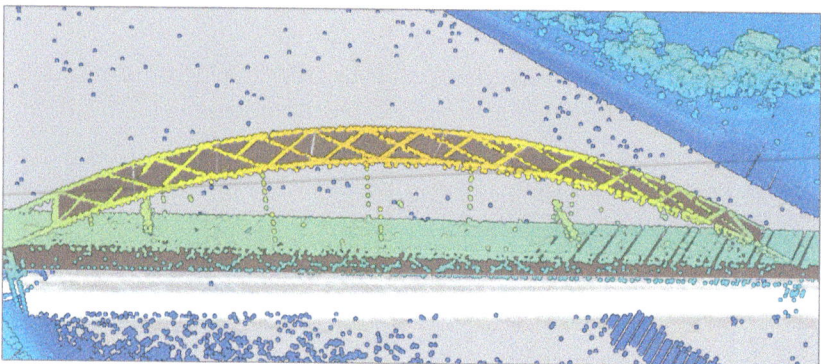

7. Turn off the **Bridges** layer and save your project.

Conduct a line-of-sight analysis

You can use lidar data to determine line-of-sight obstructions between observer points. This information can be useful for security or development purposes. ArcGIS Pro users have many 3D tools for conducting visibility studies. You will use two, **Construct Sight Lines** and **Line of Sight**. In this section, you will use two observer points already created, one from the top of the US Steel building (**Observer1**) and the other at the fountain at Point State Park (**Observer2**).

1. Zoom to the **Line of Sight View** bookmark for the scene and turn on the **Observer1** and **Observer2** layers.

2. Search for and open the **Construct Sight Lines** tool. Apply these settings and run the tool:
    - **Observer Points**: Observer1
    - **Target Features**: Observer2
    - **Output**: SightLine

    A faint sight line appears in the view between the two observer points.

3. Search for and open the **Line of Sight** tool. Apply these settings and run the tool:
    - **Input Surface**: Chapter11LasDataset.lasd
    - **Input Line Features**: SightLine
    - **Output Feature Class**: LineOfSight

4. Turn off the **Chapter11LasDataset.lasd** layer to see the features that are visible (green) and not visible (red) for **Observer1**, on top of the building.

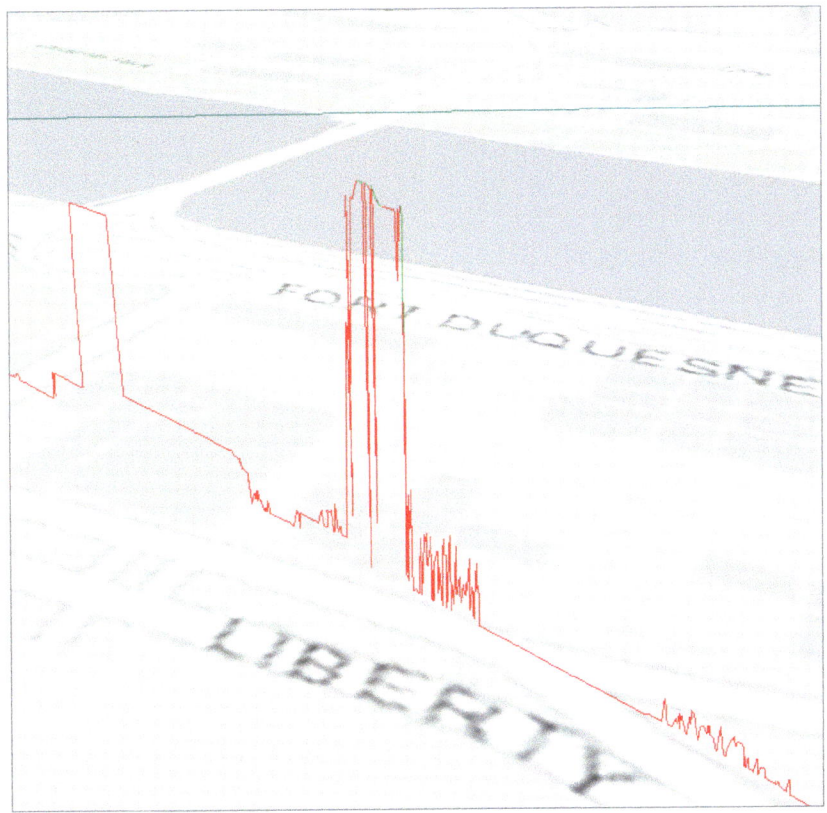

YOUR TURN

Turn on the **Bldgs** layer and symbolize with **80** percent transparency to better see the obstructions. Save your project.

Tutorial 11-5: Working with 3D features

This tutorial will explore a few of the many ArcGIS Pro 3D edit and create tools. Some edit functions work only with features that are z-enabled. If your features are not created as 3D features, you must convert them to 3D before editing using the geoprocessing tools in the **3D Analyst** toolbox. In this tutorial, you will edit building polygons that are already 3D features to create multiple floors in a building, view floors using a range slider, and manually edit polygons' heights using z-constraints.

Extrude floors

The buildings you will edit are the Allegheny County Courthouse and the old county jail, designed by the architect H. H. Richardson.

1. Open **Tutorial11-5.aprx** from the **Chapter11\Tutorials** folder and save it as **Tutorial11-5YourName.aprx**.

   First, you will create 3D floors for both the courthouse and the jail using the **Duplicate Vertical** tool. You can also use this tool to copy points or lines (for example, furniture or pipes) in a positive or negative direction if your features are aboveground or belowground. You can also select and sketch on each new floor polygon.

2. On the **Edit** tab, in the **Tools** gallery, click the **Duplicate Vertical** button.

3. Click the **Allegheny County Courthouse** polygon (the building on the left) and make these changes in the **Modify Features** pane:
   - **Vertical Offset:** 20 ft
   - **Number of times to be duplicated:** 4

4. Click **Duplicate**.

   There are now five floors.

5. Clear the selected features and save your edits.

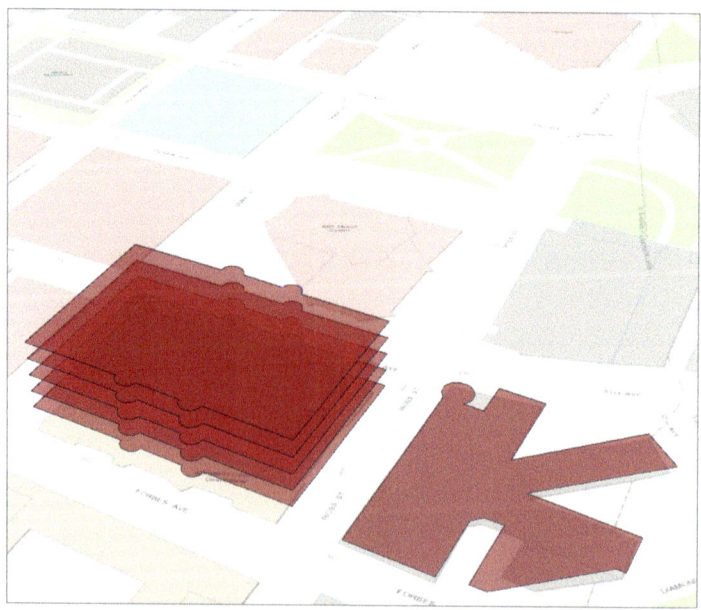

> **YOUR TURN**
>
> Use the **Duplicate Vertical** tool to extrude the floors of the old jail (the building on the right), set an offset of **20** feet, and use **3** for the number of times to duplicate. Clear your selections and save your edits.

Use a range slider to view building floors

Setting range values offers a way to visualize certain floors in a building. This visualization method is especially useful if a floor contains detailed information or a building has many floors. You can use this tool to visualize numeric values in an attribute table, including property values for parcels, crimes in neighborhoods, and so on.

1. Zoom to the **Building Floors** bookmark.

2. Open the **Courthouse3D** attribute table and sort the **Name** field in ascending order.

3. Select each floor of the courthouse, and in the corresponding **FloorNumber** field, type **1** for the first (lowest) floor, **2** for the second floor, **3** for the third floor, and so on. Repeat for the jail floors.

Name	FloorNumber
Courthouse	1
Courthouse	2
Courthouse	3
Courthouse	4
Courthouse	5
Jail	1
Jail	2
Jail	3
Jail	4

4. Clear your selection, save your edits, and close the table.

5. In the **Contents** pane, right-click **Courthouse3D** and click **Properties**. On the **Range** tab, click **Add Range**. For **Start Field**, make sure **FloorNumber** is selected, click **Add**, and click **OK**.

    A slider bar is added.

6. Starting at the bottom, drag the slider up to **3** and notice that floors **1** and **2** disappear.

   You can use range sliders in 3D animations and modify range properties on the **Range** tab in the map.

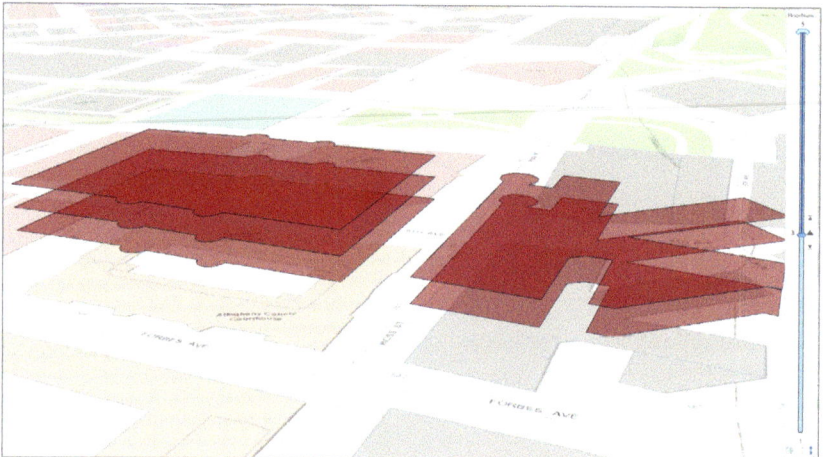

7. Drag the slider down to **1** to view all floors.

8. On the ribbon, click the **Range** tab, and in the **Active Range** group, change the **Name** value to **<None>**.

   This will turn off the range slider. You can also permanently remove the range by clicking **Courthouse** > **Properties** > **Range**.

Edit a building's height using dynamic constraints and the attribute table

Buildings sometimes consist of multiple polygons at different heights. If these heights are not already derived from lidar data, you can use interactive handles to adjust the building height dynamically using a z-constraint or by typing the building height using attributes.

1. In the **Contents** pane, turn off **Courthouse3D** and turn on **Courthouse3DTowers**.

2. On the **Edit** tab, click **Modify**, and in the **Modify Features** pane, under **Alignment**, click **Scale**.

3. Click the large tower (the building on the left) on Grant Street.

The dynamic constraint icon will appear on the tower polygon. If your icon does not appear, you can click **Project** > **Options** > **Editing** and turn on **Show Dynamic Constraints in the Map**. You can adjust the map if necessary to better see the tower and constraint icon.

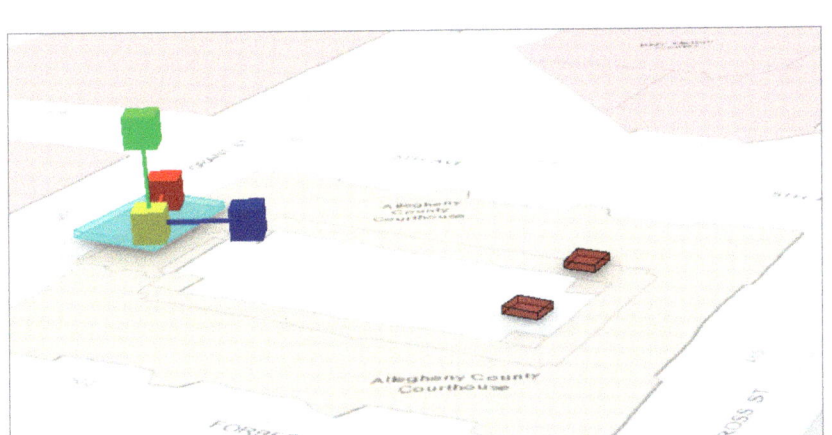

4. Click the green (**Z**) constraint to scale the tower in the z-direction.

   If you had lidar data, you could snap to those points to determine the building height.

5. Click to finish the tower approximately at the height shown.

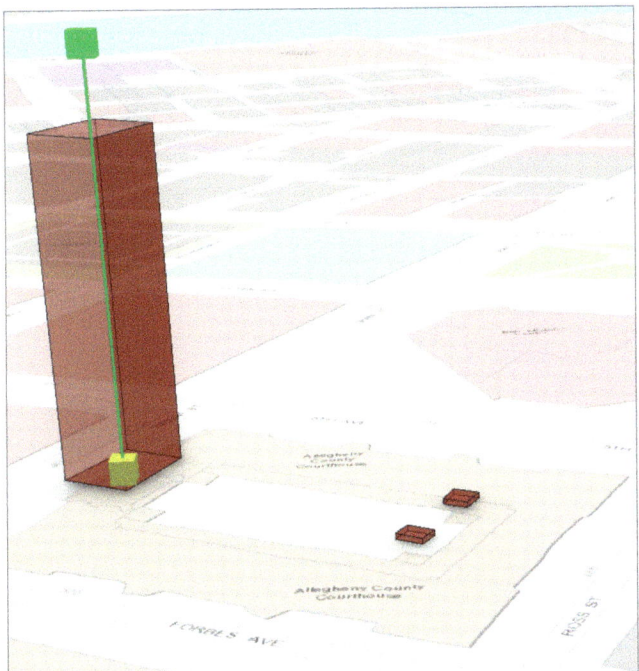

6. Save your edits.

> **YOUR TURN**
>
> An alternative to dragging features is to type the building heights in the corresponding attribute table. Open the **Courthouse3DTowers** attribute table, and for **Tower 1** and **Tower 2**, set their **Height** to **150**. Clear your selected features and save your edits and your project.

Tutorial 11-6: Using procedural rules and multipatch models

A CityEngine rule package (.rpk) is a file that contains a compiled rule and all the assets (textures and 3D models) that the rule logic uses for creating 3D content. You can use these packages in ArcGIS Pro to create symbology that constructs and draws the procedural features on the fly from the source data. Another method creates 3D models and stores them as a feature class called a multipatch, whose features are a collection of patches that represent the boundary of a 3D object. A multipatch stores color, texture, transparency, and geometric data in its features.

Here, you will apply a predefined rule package to Pittsburgh's tallest office building, the US Steel building. You will also view multipatch features whose building facades were created in CityEngine using actual building facades. These features can take a long time to render, so it's recommended that you use a high-end graphics card and follow the hardware requirements.

Apply building rules using stacked blocks

When you apply procedural rules, you must display features as layers in a scene. The feature class polygon itself does not have to include z-values, but it must be in a scene, and you can use 2D layers, such as building polygons.

1. Open **Tutorial11-6.aprx** from the **Chapter11\Tutorials** folder and save it as **Tutorial11-6YourName.aprx**.

   The project opens with a scene—US Steel Building—and one 2D building polygon footprint.

   You can apply a procedural rule to a building for a stacked block or more realistic high-rise or office building.

2. In the **Contents** pane, click the red symbol for **USSteelBldg**.

3. In the **Symbology** pane, on the **Gallery** tab, in the search box, type **Procedural** and press **Enter**.

   Procedural symbols will update with new software releases and can be downloaded and added from CityEngine or ArcGIS Living Atlas.

4. Click the **Stacked Blocks** procedural symbol.

5. In the **Symbology** pane, on the **Properties** tab, click the **Layers** button.

6. For **Units**, select **Feet**, and for **Levels**, type **64**.

7. For **Total Height**, click the **Container** button on the right of the current height. Select the **Height** field and click **OK**.

   The building height is set to a height field in which building heights are already entered.

8. Click **Apply** and drag the **USSteelBldg** layer to **3D Layers**.

9. In the **Contents** pane, right-click **USSteelBldg** and click **Properties**. In the **Layer Properties** dialog box, click the **Elevation** tab, and for **Features are**, select **On the ground** and click **OK**.

The building will now be wrapped, showing the number of levels, or floors.

Apply an international building rule

The unit for international buildings is meters, so you will type the height instead of using the building height field, which is in feet.

1. In the **Symbology** pane, on the **Gallery** tab, type **Procedural** and click the **International Building** symbol.

2. Click the **Properties** tab and click the **Layers** button.

3. Type or make these selections and click **Apply**:
    - **Units**: Meters
    - **BuildingType**: HighRise
    - **FloorHeightGround**: 7
    - **FloorHeightUpper**: 7
    - **TotalHeight**: 250

Although the result does not look exactly like the actual US Steel building, it is more realistic than a wrapped-level building.

View multipatch models of buildings and street furniture

Smithfield Street is a study area in downtown Pittsburgh using multipatch layers of exported SketchUp Collada (.dae) files with realistic building images and street furniture. You will turn on these layers to explore multipatch models.

1. Open the **Smithfield Street** scene.

   Wait while the building facades display with textures.

2. Turn on the **Smithfield Street Furniture** layer, change the basemap to **Dark Gray Canvas**, and wait for the view to render.

This layer of detailed features, such as planters, garbage and recycling cans, and newsstands, may take a while to render, depending on your graphics card.

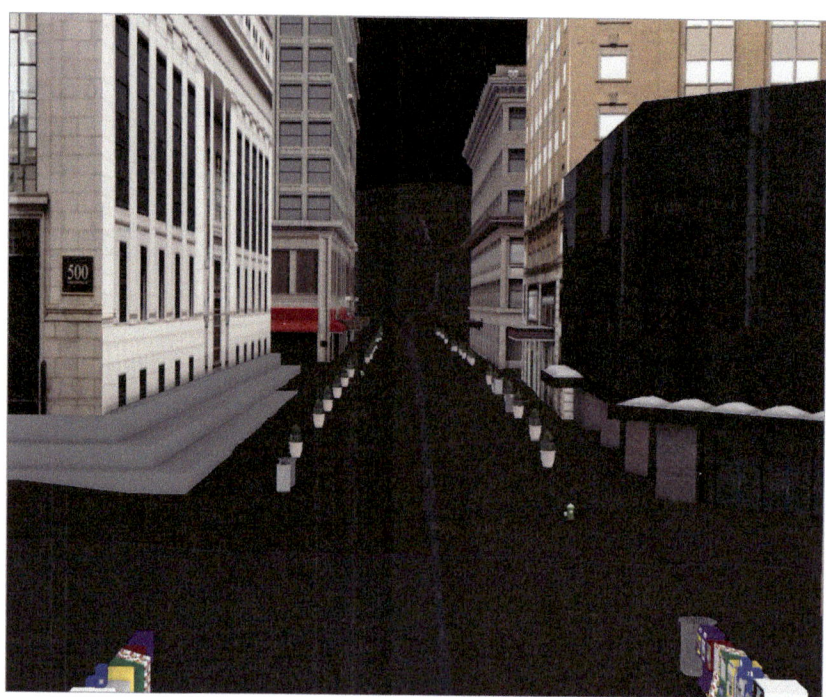

YOUR TURN

Turn off the **Smithfield Street Furniture** layer and turn on the **SmithfieldTextured2** layer. View the scene from various locations, including the **Smithfield Street Bridge** bookmark. Turn on the **Smithfield Buildings** layer, and on the **Feature Layer** tab, in the **Effects** group, change the **Transparency** to **10.0**%. Save your project.

Tutorial 11-7: Creating an animation

Animations are created by capturing an ordered set of viewpoints as keyframes and managing how the camera transitions between them. In this tutorial, you will take advantage of the bookmarks already in the scene to build a fly-through animation of downtown Pittsburgh. You will then improve the flight path and flight speed by manually inserting more keyframes and adjusting their timing.

Add an animation to the project and create keyframes

1. Open **Tutorial11-7.aprx** from the **Chapter11\Tutorials** folder and save it as **Tutorial11-7YourName.aprx**.

2. On the **Map** tab, click **Bookmarks** > **Manage Bookmarks**, and in the **Bookmarks** pane, double-click each bookmark in order, from **Frame1** through **Frame8**.

    You will use these bookmarks to create the animation.

3. Double-click the **Frame1** bookmark.

    This bookmark will be the first camera location of your animation.

    Adding animation to a project enables the various animation functions. As soon as you add an animation, you begin making keyframes manually. You can also import keyframes into an animation on the **Animation** tab, in the **Create** group, using the **Import** function.

4. On the **View** tab, in the **Animation** group, click the **Add** button.

    The **Animation** tab appears, and an **Animation Timeline** pane appears at the bottom of the screen. You are ready to make your first keyframe using bookmarks.

5. In the **Animation Timeline** pane, click **Create first keyframe**.

    A thumbnail image of the starting location (**Frame1** bookmark and the first keyframe) appears.

6. In the **Bookmarks** pane, double-click **Frame2**, and in the **Animation Timeline**, click the **Append next keyframe** button.

    This step adds the next keyframe to the animation.

7. Repeat step 6 for each of the remaining bookmarks, **Frame3** through **Frame8**.

8. Close the **Bookmarks** pane.

Now all the keyframes are added to the animation.

Play an animation and change the duration

The keyframes have durations of three seconds between each frame, with a total duration of 21 seconds. You can play the animation for this duration or extend the playback time.

1. On the **Animation** tab, in the **Playback** group, click the **Play** button.

   On the **Animation Timeline** tab, the spacebar is a keyboard shortcut for **Play/Pause**.

2. On the **Animation** tab, in the **Duration** box, type **00:30.000**.

3. Play the animation from the beginning.

   The animation will now play for 30 seconds.

Create a pause

Holding a keyframe will pause the animation at the selected frame. You will add a hold to create a slight pause between frames 5 and 6 along Smithfield Street Bridge.

1. On the **Animation Timeline**, double-click keyframe **5** and click the **Hold** button.

2. Play the animation from the beginning and observe the slight pause between frames 5 and 6.

Add and delete keyframes

You can move keyframes to adjust the speed of animations and insert frames between keyframes using different camera locations. You will add a new keyframe and manually adjust the camera.

1. On the **Animation Timeline**, drag the **Time Indicator** (red vertical bar) to approximately 21 seconds.

Moving the **Time Indicator** scrubs through time, and the location you move it to is where you will adjust the camera and insert a new keyframe.

2. Using the mouse wheel or keyboard keys, adjust the view to a slightly lower focal point.

3. On the **Animation Timeline**, click the **Insert** button.

   Because no keyframe existed at 21 seconds, when you click update, it will insert a new keyframe to capture the camera change and add a new keyframe using the current camera location.

4. Play the animation from the beginning.

Create a movie from the animation

Now that you have created an interesting animation, it's time to share the animation. To share the animation, you will export a movie to a file. You have several options, including exporting your movie directly to YouTube, Vimeo, and so on, or as a draft animation. You can change the file location and movie resolution, type, size, and so on in the **File Export** and **Advanced Movie** settings.

1. On the **Animation** tab, in the **Export** group, click the **Movie** button.

2. In the **Export Movie** pane, in the **Movie Export Presets** group, click the **Draft** button.

   This step creates a smaller file. The resulting quality of the file isn't high, but it is much faster to produce.

3. Under **File Name**, click **Browse**, browse to **Chapter11\Tutorials**, and type **Animation3D** as the movie name.

4. In the **Export Movie** pane, click **Export**. Wait while the movie is created.

   If you use other settings with higher resolutions or a larger size, the movie can take a long time to render. You can save the media format as separate JPG files that you can later stitch together using another animation software application.

5. In **File Explorer**, browse to **Chapter 11\Tutorials** and double-click **Chapter11Animation3D.mp4** to play the movie.

6. Save your project.

Assignments

This chapter has assignments to complete that you can download with data from ArcGIS Online at links.esri.com/GISTforPro3.4Assignments.

Data source credits

Chapter 1
\Tutorials\Chapter1.gdb\Tracts, courtesy of US Census Bureau.
\Tutorials\Chapter1.gdb\AlleghenyCounty, courtesy of US Census Bureau.
\Tutorials\Chapter1.gdb\FQHCClinics, courtesy of W. L. Gorr, Carnegie Mellon University.
\Tutorials\Chapter1.gdb\FQHCBuffer, courtesy of W. L. Gorr, Carnegie Mellon University.
\Tutorials\Chapter1.gdb\UrgentCareClinics, courtesy of W. L. Gorr, Carnegie Mellon University.
\Tutorials\Chapter1.gdb\UrgentCareClinicsBuffer, courtesy of W. L. Gorr, Carnegie Mellon University.
\Tutorials\Chapter1.gdb\Municipalities, courtesy of US Census Bureau.
\Tutorials\Chapter1.gdb\Parks, courtesy of Southwestern Pennsylvania Commission.
\Tutorials\Chapter1.gdb\Pittsburgh, courtesy of US Census Bureau.
\Tutorials\Chapter1.gdb\Population Density, courtesy of W. L. Gorr, Carnegie Mellon University.
\Tutorials\Chapter1.gdb\PovertyRiskArea, courtesy of W. L. Gorr, Carnegie Mellon University.
\Tutorials\Chapter1.gdb\PovertyDensity, courtesy of W. L. Gorr, Carnegie Mellon University.
\Tutorials\Chapter1.gdb\Rivers, courtesy of US Census Bureau.
\Tutorials\Chapter1.gdb\Streets, courtesy of US Census Bureau, TIGER.

Chapter 2
\Tutorials\Chapter2.gdb\Boroughs, courtesy of Department of City Planning, New York City.
\Tutorials\Chapter2.gdb\Facilities, courtesy of Department of City Planning, New York City.
\Tutorials\Chapter2.gdb\FireCompanies, courtesy of Department of City Planning, New York City.
\Tutorials\Chapter2.gdb\FireHouses, courtesy of Department of City Planning, New York City.
\Tutorials\Chapter2.gdb\ManhattanStreets, courtesy of Department of City Planning, New York City.
\Tutorials\Chapter2.gdb\Neighborhoods, courtesy of Department of City Planning, New York City.
\Tutorials\Chapter2.gdb\PolicePrecincts, courtesy of Department of City Planning, New York City.
\Tutorials\Chapter2.gdb\PoliceStations, courtesy of Department of City Planning, New York City.
\Tutorials\Chapter2.gdb\Water, courtesy of Department of City Planning, New York City.
\Tutorials\Chapter2.gdb\ZoningLandUse, courtesy of Department of City Planning, New York City.

Chapter 3

\Tutorials\Chapter3.gdb\MetroArtsPoints, courtesy of Bureau of Labor Statistics.
\Tutorials\Chapter3.gdb\MetroCOLIPoints, courtesy of the Council for Community and Economic Research.
\Tutorials\Chapter3.gdb\Neighborhoods, courtesy of Department of City Planning, City of Pittsburgh.
\Tutorials\Chapter3.gdb\PghStreets, courtesy of US Census Bureau.
\Tutorials\Chapter3.gdb\Pittsburgh, courtesy of US Census Bureau.
\Tutorials\Chapter3.gdb\Rivers, courtesy of US Census Bureau.
\Tutorials\Chapter3.gdb\USStates, courtesy of US Census Bureau, the Council for Community and Economic Research, Bureau of Labor Statistics.
\Tutorials\Chapter3.gdb\USStatesPoints, courtesy of US Census Bureau, the Council for Community and Economic Research, Bureau of Labor Statistics.

Chapter 4

\\Data\MaricopaCounty\PopYouth.csv, courtesy of US Census Bureau.
\Data\MaricopaCounty\Municipalities.shp, courtesy of US Census Bureau.
\Data\MaricopaCounty\MaricopaCounty.shp, courtesy of US Census Bureau.
\Data\MaricopaCounty\Tracts.shp, courtesy of US Census Bureau.
\Data\Crime.gdb\PittsburghSeriousCrimesSummer2015.shp, courtesy of City of Pittsburgh Police Bureau.
\Data\Crime.gdb\Burglaries, courtesy of City of Pittsburgh Police Bureau.
\Data\Crime.gdb\CrimeOffenses, courtesy of City of Pittsburgh Police Bureau.
\Data\Crime.gdb\Neighborhoods, courtesy of Department of City Planning, City of Pittsburgh.
\Data\Crime.gdb\Streets, courtesy of US Census Bureau.

Chapter 5

\Data\NewYorkCity\CouncilDistricts.shp, courtesy of Department of City Planning, New York City.
\Data\NewYorkCity\Libraries.dbf, courtesy of Department of City Planning, New York City.
\Tutorials\Chapter5.gdb\Counties, from ArcGIS Data and Maps (2010), courtesy of ArcUSA, US Census Bureau.
\Tutorials\Chapter5.gdb\Country, from ArcGIS Data and Maps (2004), courtesy of *ArcWorld Supplement*.
\Tutorials\Chapter5.gdb\HennepinCounty, courtesy of US Census Bureau.
\Tutorials\Chapter5.gdb\Municipalities, courtesy of Southwestern Pennsylvania Commission.
\Tutorials\Chapter5.gdb\Ocean, from ArcGIS Data and Maps, courtesy of Esri.
\Tutorials\Chapter5.gdb\Parks, courtesy of Southwestern Pennsylvania Commission.
\Tutorials\Chapter5.gdb\Tracts, courtesy of US Census Bureau.
\Tutorials\Chapter5.gdb\States, from ArcGIS Data and Maps (2010), courtesy of ArcUSA, US Census Bureau.

Chapter 6

\Tutorials\Chapter6.gdb\Boroughs, courtesy of Department of City Planning, New York City.
\Tutorials\Chapter6.gdb\BronxWater, courtesy of US Census Bureau.

\Tutorials\Chapter6.gdb\BronxWaterfrontParks, courtesy of Department of City Planning, New York City.
\Tutorials\Chapter6.gdb\BrooklynWater, courtesy of US Census Bureau.
\Tutorials\Chapter6.gdb\BrooklynLandUse, courtesy of Department of City Planning, New York City.
\Tutorials\Chapter6.gdb\BrooklynWaterfrontParks, courtesy of Department of City Planning, New York City.
\Tutorials\Chapter6.gdb\BrooklynNeighborhoods, courtesy of US Census Bureau.
\Tutorials\Chapter6.gdb\EMSFacilities, courtesy of Department of City Planning, New York City.
\Tutorials\Chapter6.gdb\FireCompanies, courtesy of Department of City Planning, New York City.
\Tutorials\Chapter6.gdb\FireHouses, courtesy of Department of City Planning, New York City.
\Tutorials\Chapter6.gdb\ManhattanBlockGroups, courtesy of US Census Bureau.
\Tutorials\Chapter6.gdb\ManhattanFireCompanies, courtesy of Department of City Planning, New York City.
\Tutorials\Chapter6.gdb\ManhattanLandUse, courtesy of Department of City Planning, New York City.
\Tutorials\Chapter6.gdb\ManhattanStreets, courtesy of US Census Bureau.
\Tutorials\Chapter6.gdb\ManhattanTracts, courtesy of US Census Bureau.
\Tutorials\Chapter6.gdb\ManhattanWater, courtesy of US Census Bureau.
\Tutorials\Chapter6.gdb\ManhattanWaterfrontParks, courtesy of Department of City Planning, New York City.
\Tutorials\Chapter6.gdb\NeighborhoodsZoningLandUse, courtesy of Department of City Planning, New York City.
\Tutorials\Chapter6.gdb\NYCBlockGroups, courtesy of US Census Bureau.
\Tutorials\Chapter6.gdb\NYCNeighborhoods, courtesy of US Census Bureau.
\Tutorials\Chapter6.gdb\NYCWaterfrontParks, courtesy of Department of City Planning, New York City.
\Tutorials\Chapter6.gdb\PoliceStations, courtesy of Department of City Planning, New York City.
\Tutorials\Chapter6.gdb\QueensNeighborhoods, courtesy of US Census Bureau.
\Tutorials\Chapter6.gdb\QueensLandUse, courtesy of Department of City Planning, New York City.
\Tutorials\Chapter6.gdb\QueensWater, courtesy of US Census Bureau.
\Tutorials\Chapter6.gdb\QueensWaterfrontParks, courtesy of Department of City Planning, New York City.
\Tutorials\Chapter6.gdb\StatenIslandWater, courtesy of US Census Bureau.
\Tutorials\Chapter6.gdb\StatenIslandWaterfrontParks, courtesy of Department of City Planning, New York City.
\Tutorials\Chapter6.gdb\UpperWestSideFireCompanies, courtesy of Department of City Planning, New York City.
\Tutorials\Chapter6.gdb\UpperWestSideTracts, courtesy of US Census Bureau.
\Tutorials\Chapter6.gdb\UpperWestSideZoningLandUse, courtesy of Department of City Planning, New York City.

\Tutorials\Chapter6.gdb\ZoningLandUse, courtesy of Department of City Planning, New York City.

Chapter 7

\Data\HBH1.dwg, courtesy of Carnegie Mellon University.
\Tutorials\Chapter7.gdb\Bldgs, courtesy of Department of City Planning, City of Pittsburgh.
\Tutorials\Chapter7.gdb\BldgsOriginal, courtesy of Department of City Planning, City of Pittsburgh.
\Tutorials\Chapter7.gdb\BusStopCrossWalk, courtesy of Kristen Kurland, Carnegie Mellon University.
\Tutorials\Chapter7.gdb\Greenspaces (from Parks), courtesy of Southwestern Pennsylvania Commission.
\Tutorials\Chapter7.gdb\Streets, courtesy of US Census Bureau.
\Tutorials\Chapter7.gdb\MainCampusStudyAreaBldgs, courtesy of Department of City Planning, City of Pittsburgh.
\Tutorials\Chapter7.gdb\StudyAreaBldgs, courtesy of Department of City Planning, City of Pittsburgh.
\Tutorials\Chapter7.gdb\Water, courtesy of US Census Bureau.

Chapter 8

\Data\AliasTable.csv, courtesy of W. L. Gorr, Carnegie Mellon University.
\Data\AttendeesAlleghenyCounty.csv, courtesy of FLUX.
\Data\AttendeesAlleghenyPARegion.csv, courtesy of FLUX.
\Data\Clients.csv, courtesy of Kristen Kurland, Carnegie Mellon University.
\Data\GroceryStores.csv, courtesy of W. L. Gorr, Carnegie Mellon University.
\Tutorials\Chapter8.gdb\AlleghenyCounty.AllCoZIP, courtesy of US Census Bureau, TIGER.
\Tutorials\Chapter8.gdb\AlleghenyCounty.AlleghenyCounty, courtesy of US Census Bureau, TIGER.
\Tutorials\Chapter8.gdb\AlleghenyCounty.Municipalities, courtesy of US Census Bureau, TIGER.
\Tutorials\Chapter8.gdb\Streets, courtesy of US Census Bureau, TIGER.
\Tutorials\Chapter8.gdb\PARegion.PARegion, courtesy of US Census Bureau, TIGER.
\Tutorials\Chapter8.gdb\PARegion.PARegionZIP, courtesy of US Census Bureau, TIGER.
\Tutorials\Chapter8.gdb\PittsburghCBD.CBDOutline, courtesy of Department of City Planning, City of Pittsburgh.
\Tutorials\Chapter8.gdb\PittsburghCBD.CBDStreets, courtesy of Department of City Planning, City of Pittsburgh.
\Tutorials\Chapter8.gdb\USStates, courtesy of US Census Bureau, TIGER.

Chapter 9

\Data\Exponential.xlsx, courtesy of W. L. Gorr, Carnegie Mellon University.
\Data\PittsburghNetworkDataset.gdb\PittsburghStreets\PittsburghStreets, courtesy of US Census Bureau, TIGER.
\Data\PittsburghNetworkDataset.gdb\PittsburghStreets\PittsburghStreets_ND, courtesy of US Census Bureau, TIGER, and W. L. Gorr, Carnegie Mellon University.

\Data\PittsburghNetworkDataset.gdb\PittsburghStreets\PittsburghStreets_ND_Junctions, courtesy of US Census Bureau, TIGER, and W. L. Gorr, Carnegie Mellon University.
\Tutorials\Chapter9.gdb\DrugViolations, courtesy of City of Pittsburgh Police Bureau.
\Tutorials\Chapter9.gdb\Neighborhoods, courtesy of Department of City Planning, City of Pittsburgh.
\Tutorials\Chapter9.gdb\Pittsburgh, courtesy of US Census Bureau, TIGER.
\Tutorials\Chapter9.gdb\PittsburghBlockCentroids, courtesy of US Census Bureau, TIGER.
\Tutorials\Chapter9.gdb\PoliceZones, courtesy of Department of City Planning, City of Pittsburgh.
\Tutorials\Chapter9.gdb\Pools, courtesy of Pittsburgh CitiParks Department.
\Tutorials\Chapter9.gdb\Pooltags, courtesy of Pittsburgh CitiParks Department.
\Tutorials\Chapter9.gdb\PovertyRiskContour, courtesy of US Census Bureau, TIGER, and W. L. Gorr, Carnegie Mellon University.
\Tutorials\Chapter9.gdb\Rivers, courtesy of US Census Bureau, TIGER.
\Tutorials\Chapter9.gdb\Schools, courtesy of Department of City Planning, City of Pittsburgh.
\Tutorials\Chapter9.gdb\SeriousViolentCrimes, courtesy of City of Pittsburgh Police Bureau.

Chapter 10

\Data\LandUse_Pgh.tif, image courtesy of US Geological Survey, Department of the Interior/USGS.
\Data\LandUse.lyr, courtesy of US Geological Survey, Department of the Interior/USGS, and W. L. Gorr, Carnegie Mellon University.
\Tutorials\Chapter10.gdb\NED, courtesy of US Geological Survey, Department of the Interior/USGS.
\Tutorials\Chapter10.gdb\Neighborhoods, courtesy of Department of City Planning, City of Pittsburgh.
\Tutorials\Chapter10.gdb\OHCA, courtesy of Children's Hospital of Pittsburgh.
\Tutorials\Chapter10.gdb\Pittsburgh, courtesy of US Census Bureau, TIGER.
\Tutorials\Chapter10.gdb\PittsburghBlkGrps, courtesy of US Census Bureau, TIGER.
\Tutorials\Chapter10.gdb\Rivers, courtesy of US Census Bureau, TIGER.
\Tutorials\Chapter10.gdb\ZoningCommercialBuffer, courtesy of Department of City Planning, City of Pittsburgh, and W. L. Gorr, Carnegie Mellon University.

Chapter 11

\Data\LASFiles\1336704E409152N, courtesy of Pictometry International Corp.
\Data\LASFiles\1336704E411792N, courtesy of Pictometry International Corp.
\Data\LASFiles\1339344E409152N, courtesy of Pictometry International Corp.
\Data\LASFiles\1339344E411792N, courtesy of Pictometry International Corp.
\Data\LASFiles\1341984E409152N, courtesy of Pictometry International Corp.
\Data\LASFiles\1341984E411792N, courtesy of Pictometry International Corp.
\Data\PACPIT14_LiDAR_Delivery_Area.shp, courtesy of Pictometry International Corp.
\Tutorials\Chapter11.gdb\AnimationBldgs, courtesy of Department of City Planning, City of Pittsburgh.
\Tutorials\Chapter11.gdb\Bldgs, courtesy of Department of City Planning, City of Pittsburgh.
\Tutorials\Chapter11.gdb\Bridges, courtesy of Department of City Planning, City of Pittsburgh.

\Tutorials\Chapter11.gdb\Contours, courtesy of Department of City Planning, City of Pittsburgh.

\Tutorials\Chapter11.gdb\Courthouse3D, courtesy of Department of City Planning, City of Pittsburgh.

\Tutorials\Chapter11.gdb\Courthouse3DTowers, courtesy of Southwestern Pennsylvania Commission.

\Tutorials\Chapter11.gdb\Parks, courtesy of Southwestern Pennsylvania Commission.

\Tutorials\Chapter11.gdb\ParkTrees, courtesy of Department of City Planning, City of Pittsburgh.

\Tutorials\Chapter11.gdb\Rivers, courtesy of Southwestern Pennsylvania Commission.

\Tutorials\Chapter11.gdb\SmithfieldBldgs, courtesy of Department of City Planning, City of Pittsburgh.

\Tutorials\Chapter11.gdb\SmithfieldFurniture, courtesy of Department of City Planning, City of Pittsburgh.

\Tutorials\Chapter11.gdb\SmithfieldTextured1, courtesy of Kristen Kurland, Carnegie Mellon University.

\Tutorials\Chapter11.gdb\SmithfieldTextured2, courtesy of Kristen Kurland, Carnegie Mellon University.

\Tutorials\Chapter11.gdb\StreetCurbs, courtesy of Department of City Planning, City of Pittsburgh.

\Tutorials\Chapter11.gdb\StreetTrees, courtesy of Department of City Planning, City of Pittsburgh.

\Tutorials\Chapter11.gdb\StudyArea (selected features from PACPIT14_LiDAR_Delivery_Area), courtesy of Pictometry International Corp.

\Tutorials\Chapter11.gdb\USSteelBldg, courtesy of Department of City Planning, City of Pittsburgh.

About Esri Press

Esri Press is an American book publisher and part of Esri, the global leader in GIS software, location intelligence, and mapping. Since 1969, Esri has supported customers with geographic science and geospatial analytics, what we call The Science of Where®. We take a geographic approach to problem-solving, brought to life by modern GIS technology, and are committed to using science and technology to build a sustainable world.

At Esri Press, our mission is to inform, inspire, and teach professionals, students, educators, and the public about GIS by developing print and digital publications. Our goal is to increase the adoption of ArcGIS and to support the vision and brand of Esri. We strive to be the leader in publishing great GIS books, and we are dedicated to improving the work and lives of our global community of users, authors, and colleagues.

Acquisitions
Stacy Krieg
Claudia Naber
Alycia Tornetta
Jenefer Shute

Product Engineering
Craig Carpenter
Maryam Mafuri

Editorial
Carolyn Schatz
Mark Henry
David Oberman

Production
Monica McGregor
Victoria Roberts

Sales & Marketing
Eric Kettunen
Sasha Gallardo
Beth Bauler

Contributors
Christian Harder
Matt Artz

Business
Catherine Ortiz
Jon Carter
Jason Childs

Related titles

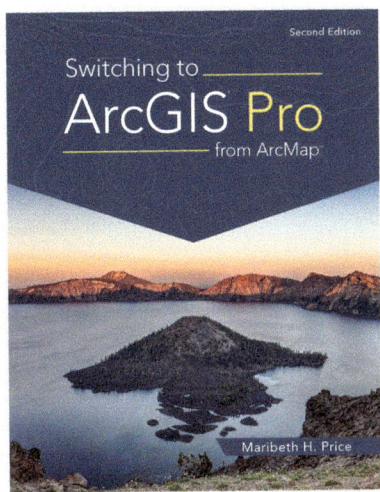

Switching to ArcGIS Pro from ArcMap
Maribeth H. Price
9781589487314

Top 20 Essential Skills for ArcGIS Pro
Bonnie Shrewsbury and Barry Waite
9781589487505

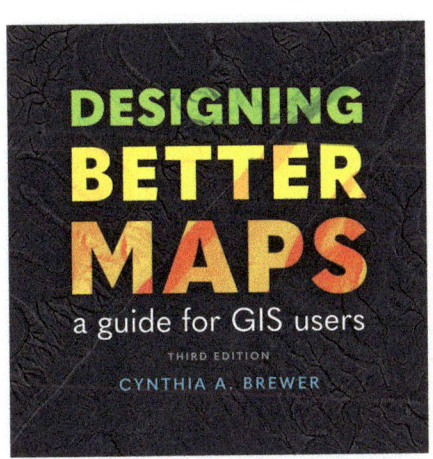

Designing Better Maps
A Guide for GIS Users, third edition
Cynthia A. Brewer
9781589487826

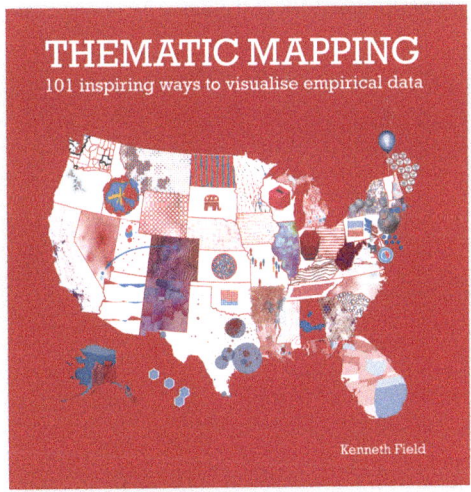

Thematic Mapping
101 Inspiring Ways to Visualise Empirical Data
Kenneth Field
9781589485570

For information on Esri Press books, e-books, and resources, visit our website at
esripress.com.